THE HEIGHTS

ANATOMY OF A SKYSCRAPER

Kate Ascher

Art direction by Design Language

Research by Rob Vroman

The Penguin Press

New York 2011

THE PENGUIN PRESS
Published by the Penguin Group
Penguin Group (USA) Inc., 375 Hudson Street,
New York, New York 10014, U.S.A.
•
Penguin Group (Canada), 90 Eglinton Avenue East, Suite 700,
Toronto, Ontario, Canada M4P 2Y3
(a division of Pearson Penguin Canada Inc.)
•
Penguin Books Ltd, 80 Strand, London WC2R ORL, England
•
Penguin Ireland, 25 St. Stephen's Green, Dublin 2, Ireland
(a division of Penguin Books Ltd)
•
Penguin Books Australia Ltd, 250 Camberwell Road, Camberwell,
Victoria 3124, Australia
(a division of Pearson Australia Group Pty Ltd)
•
Penguin Books India Pvt Ltd, 11 Community Centre, Panchsheel Park,
New Delhi – 110 017, India
•
Penguin Group (NZ), 67 Apollo Drive, Rosedale, Auckland 0632,
New Zealand (a division of Pearson New Zealand Ltd)
•
Penguin Books (South Africa) (Pty) Ltd, 24 Sturdee Avenue,
Rosebank, Johannesburg 2196, South Africa
•
Penguin Books Ltd, Registered Offices:
80 Strand, London WC2R ORL, England

First published in 2011 by The Penguin Press,
a member of Penguin Group (USA) Inc.

1 3 5 7 9 10 8 6 4 2

Copyright © Kate Ascher, 2011
All rights reserved

Credits for contributing artists appear on page 193.
Illustrations by Design Language, Copyright © Penguin Group (USA) Inc., 2011

Photography credits appear on page 195.

ISBN 978-1-59420-303-9

Printed in China
Set in Tisa and Forza
Designed by Design Language

Jacket design: Dan Donohue and George Kokkinidis
Jacket illustration: Vic Kulihin

To Peter and Pilar

A NOTE TO READERS

The downturn in the global economy that began in 2008 did little to dent the world's appetite for tall buildings. In the two years it took to write this book, hundreds of new skyscrapers opened for business across the globe. And the race for the sky continues, with hundreds more targeted for completion in 2011 and 2012.

While many of these new skyscrapers and their older counterparts are remarkable, very few have found their way into the pages that follow. Rather than celebrate individual skyscrapers, the book focuses on the forces that shape the way tall buildings work. Particular skyscrapers are highlighted only to shed light on the concepts that underpin their form and operation.

Not surprisingly, a wealth of detail and technical data are required to explain how these remarkable structures work—and each of the ensuing chapters is comprised of both. While every effort has been made to ensure that this information is accurate at the time of publication, neither the publisher nor the author assumes responsibility for errors or for changes that occur after publication.

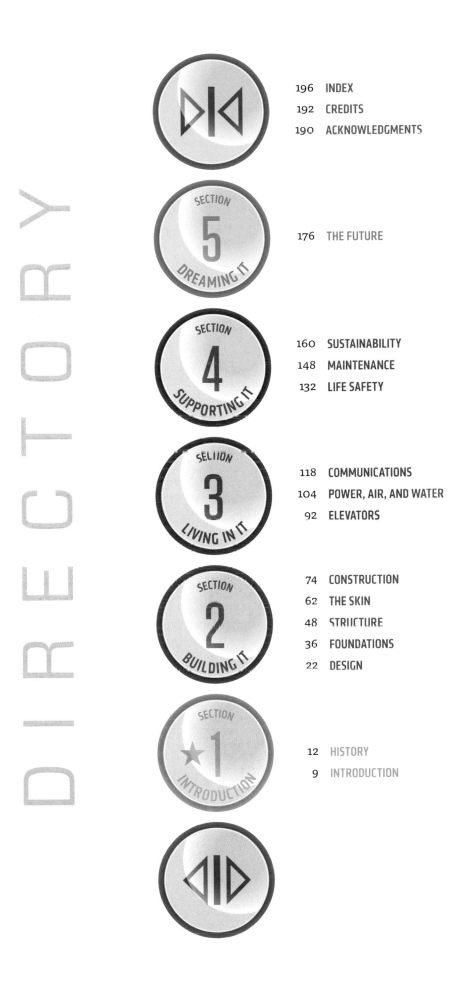

INTRODUCTION

What is the chief characteristic of the tall office building? It is lofty. It must be tall. The force and power of altitude must be in it, the glory and pride of exaltation must be in it. It must be every inch a proud and soaring thing, rising in sheer exaltation that from bottom to top it is a unit without a single dissenting line.

—Louis Sullivan, "The Tall Office Building Artistically Considered," *Lippincott's* magazine, 1896

hen it comes to buildings, size matters—more so today than ever before. Look up in the heart of any of the world's major cities and your eyes will likely alight upon a towering, glass-walled structure—if not literally scraping the sky, then certainly pointing in that direction.

The proliferation of skyscrapers is accelerating rapidly. Prior to the year 2000, fewer than 250 buildings around the world reached higher than 600 feet (180 meters); between 2000 and 2009, that number more than doubled. And it continues to grow faster than ever before: at the beginning of 2010, almost 400 new skyscrapers were under construction around the world. Not only are there more tall buildings, but they are in more places. Once a purely American phenomenon, the construction of skyscrapers is now very much a global one. Of the 38 skyscrapers over 600 feet (180 meters) completed in 2009, 22 of them were in Asia and seven were in the Middle East. The "tallest" metropolis in the world is in Asia: the combined height of Hong Kong's skyscrapers is roughly three times that of New York City's.

So prolific are these towers today in the world's metropolises, and so enthusiastic are their planners, that there are now adjectives that differentiate between them: "tall" is often used to describe skyscrapers between 500 and 1,000 feet (150 to 300 meters); anything above that is considered "supertall." Even measuring the building is now a science: do you measure it from street level or from the basement? To the highest occupied floor or to the top of its crown?

Some readers may wonder why we should care how tall these towers are or how many of them exist. Are they not just the product of speculation and greed on the part of a handful of wealthy developers? Or the modern incarnation of some age-old form of civic pride? The answer is no—they are much more than that. Thanks to a variety of factors causing people to migrate from rural to urban areas, they are increasingly the way the world's people live today—and almost certainly the way most of them will live in the future.

A United Nations study found that half of all humanity lived in urban areas in 2008—and that percentage is rising. By 2050, the study estimated a full 70 percent of the world's population will be city residents. Just how our cities will be able to accommodate these people physically is thus an increasingly important question—not just to urban planners but also to architects, transportation engineers, and sociologists.

It is not simply the external configuration of our tall buildings that we should care about. Modern society spends an estimated 90 percent of its time inside, a far cry from early civilizations whose livelihoods were tied closely to the outdoors. Today we work inside, we sleep inside, and typically we socialize inside as well. Whether we know it or not, the inside workings of our built environment affect our health and well-being—and shape almost every facet of our daily existence.

Modern skyscrapers are effectively small cities, providing infrastructure and services to thousands of inhabitants, often at enormous heights. They may be residents of apartments and hotel rooms, workers in corporate offices, or users of a more transient nature—shoppers, restaurant goers, or observation deck visitors. Regardless of their purpose aloft, these sky dwellers demand the same array of services they would require if they were closer to the ground: shelter, fresh air, clean water, plumbing, electricity, vertical transportation, and communication. Just how these services are provided thousands of feet up in the sky lies at the heart of this book.

The first section focuses on the building of skyscrapers, and on their component parts. Nowhere more than here, in the design and engineering that goes into these extraordinary structures, does a tall building differ in its needs from a smaller one. The interplay between architecture and engineering informs everything—from the building's foundations to its structure to the skin that encloses it and to the way it is constructed.

The second section of the book looks at the services that must be brought into the skyscraper once its core and shell have been completed. Over the course of history, tall structures—such as pyramids, cathedrals, and bell towers—were generally uninhabited and thus needed little by way of supporting services. In contrast, today's skyscrapers rely on a vast network of infrastructure to sustain life in the sky—from power, air, and water to telecommunications and elevators.

The third and final section of the book touches on the more mundane concerns needed to keep tall buildings functioning on a day-to-day basis: maintenance and cleaning, safety, and sustainable technologies. Less sexy perhaps than other topics, these functions touch more immediately and more intimately on the daily lives of skyscraper occupants than almost anything else.

Across all three of the sections the book draws on examples from around the world to illustrate concepts and ideas. Yet it is in no way an exhaustive study of the world's tall buildings; indeed, many interesting and architecturally significant towers receive scarcely a mention on the pages that follow. Instead, the book is designed to leave the reader with an understanding of—and hopefully a greater appreciation for—the inner workings of this new and potentially highly sustainable urban form: the vertical city.

HISTORY

uilding tall, of course, is not unique to the present age. For all of recorded history, mankind's fascination with structures that rise toward the sky has been constant. The Egyptians devised an ingenious system of moving rock to unprecedented heights as they completed the Great Pyramid of Giza. Equally impressive were the great Gothic cathedrals of France and the medieval towers of Italy—symbols of religious or personal power meant to instill awe and respect. The connection between size and power was not lost on the new nation west of the Atlantic, whose civic buildings were often as grand and as tall as its churches—and whose soaring state capitol buildings remain important symbols today.

From this lineage, then, comes the skyscraper—a modern-day symbol of commercial power and civic pride. But today's skyscrapers distinguish themselves from everything tall that has come before in two important and related respects: they are built to make money and, to do that, must support people living or working inside of them. For the first time in history, the challenge is not just to construct a monumentally tall structure but also to make every inch of it work for human beings.

The modern skyscraper is very much a global phenomenon, but its origins are almost entirely American. This is perhaps not surprising, given that its original purpose was to make money out of real estate. Its heritage is also an urban one, as skyscrapers only make sense as a commercial undertaking where the cost of land is high and rents can cover the costs of constructing and maintaining a building hundreds of feet up in the sky.

The history of skyscrapers is thus a story of urban America, and even more so of just two of its cities. Though historians disagree on the details, it is generally accepted that the form was invented in Chicago and New York in the late 1880s—and born not of lofty ambition but of more mundane, commercial necessity. Maturing American businesses

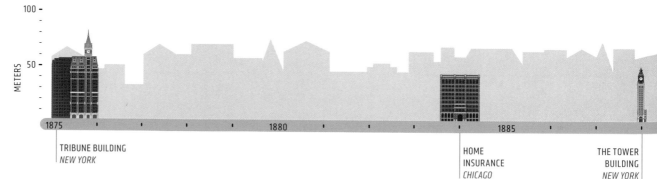

Gray skyline shows the average height of the tallest 10 buildings built each year.

METERS

100

50

1875　　　　　　　1880　　　　　　　1885

TRIBUNE BUILDING
NEW YORK

HOME INSURANCE
CHICAGO

THE TOWER BUILDING
NEW YORK

(or at least their executives) wanted to be located downtown—where high land values ensured that new structures would need to reach as high as possible to cover their costs.

In Chicago, space was at a particular premium. Much of the downtown had been gutted by a devastating fire in 1871, leaving office space scarce. Despite the increased demand, buildings could grow no taller than a dozen or so stories: as they reached higher, the load-bearing masonry walls supporting them had to get thicker—rendering the lower floors of the building less and less usable. The structural requirements of supporting the gravity load of a building put a natural cap on its height.

A series of inventions would, of course, allow buildings to rise higher. Following close on the heels of the elevator, which appeared in the 1870s, the development of the internal steel skeleton permitted larger windows and more usable floor area. By running cast-iron columns within the supporting masonry walls, William Le Baron Jenney found that the thickness of the walls of his Home Insurance Building could shrink significantly—opening up more of the building for commercial use and enabling greater light penetration. The new metal frames provided more rentable area at the base of the building, and quickly proved cheaper to construct.

The initial skyscraper skeletons were made of cast iron, but framing technology would develop quickly. As Henry Bessemer perfected his technique for the mass production of steel in the second half of the nineteenth century, the price of steel dropped dramatically—from $167 per ton in 1867 to $24 per ton in 1895. By the turn of the century, steel had replaced cast iron as the backbone of choice for new skyscrapers, and buildings of 15 to 20 stories had been completed in both New York and Chicago.

Advances in elevator technology played a large role in making these new skyscrapers habitable, with hydraulic power replacing steam in the 1870s and providing access to greater heights. Heating and cooling technology also advanced by leaps and bounds, thanks to the invention and expansion of steam heating systems, electrical plumbing pumps, and district central heating. Electricity gave skyscrapers a big boost as well; Thomas Edison's invention of the incandescent bulb made it possible for office workers to control light at their own workplace with almost no effort.

The inventions that gave rise to the skyscraper era went beyond just those providing human comfort. Steam power paved the way for the use of new and more powerful tools during the excavation and erection of buildings. Simultaneously, early experiments with telephones were under way—efforts that would ultimately transform communications between office dwellers and the outside world. In sum, the modern skyscraper was very much a product of the Industrial Revolution—and of a myriad of its most transformative inventions.

But while the Industrial Revolution was felt simultaneously across the Atlantic, in both Britain and on the continent, it never translated there into the development of tall buildings. Nothing in Europe would rise as tall as the buildings under construction in Chicago and New York. Nor would any other U.S. city come even close to competing with Chicago and New York in pushing the boundaries of commercial construction. The race for height—over an entire century in time—would be a tale of two cities.

Initially the emphasis was on improving the performance of the metal skeleton supporting the building's weight. The Tower Building in New York—completed four years after Chicago's Home Insurance Building—was the first building to rely exclusively on a steel skeleton for support, albeit one encased in masonry. The first tower to incorporate a wire-braced frame to withstand sway was begun in Chicago in 1888, sporting a rather confusing name: the Manhattan Building. Soon after, the idea of relying on a system of steel beam-based cages to support the building's weight made its debut in New York. Built by the engineer George Fuller, it was originally called the Fuller Building but—due to its irregular shape—quickly became known as the Flatiron Building.

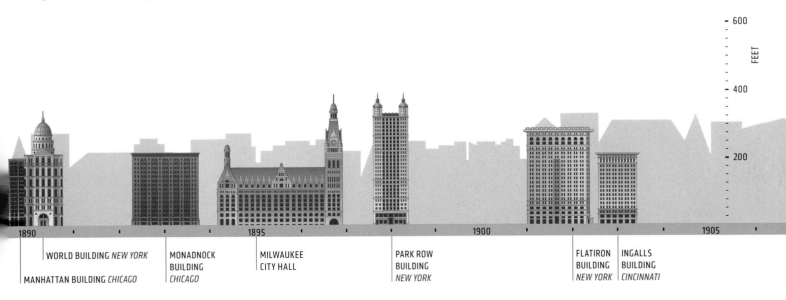

1890		1895		1900		1905

WORLD BUILDING *NEW YORK*

MANHATTAN BUILDING *CHICAGO*

MONADNOCK BUILDING *CHICAGO*

MILWAUKEE CITY HALL

PARK ROW BUILDING *NEW YORK*

FLATIRON BUILDING *NEW YORK*

INGALLS BUILDING *CINCINNATI*

Cathedrals of Commerce

Developments in Chicago and New York influenced each other tremendously at the outset of the skyscraper era, particularly with respect to engineering and structure. But the engineering advances did not always translate into pioneering heights, particularly in Chicago, where municipal laws passed in 1893 limited the height of downtown buildings to 130 feet (40 meters), or roughly 10 stories. Chicago's height regulations resulted in the development of boxy towers with relatively large footprints, penetrated at their center by courtyards to provide light.

New York, with no height cap, would very quickly produce a different skyline. To maximize value on the smaller lots that had been defined in the Commissioners' Plan of 1811, developers built taller, more slender towers. By 1913, New York had roughly 1,000 buildings over the height limit in effect in Chicago; 50 of them were more than 20 stories. New York was competing against only itself in the race for height.

But, importantly, this race for height was geared to more than just maximizing real estate profits. The egos and pride of New York's business world drove the competition for height in the early years of the twentieth century. Big corporations banked heavily on the promotional value of signature real estate. The 612-feet (187 meters) Singer Building opened in 1908 on lower Broadway; one year later, the 700-feet (213 meters) Met Life Building, located on Madison Square Park, was completed. Each of these briefly held the title of world's tallest building.

Perhaps the most famous of these corporate undertakings was F. W. Woolworth's Woolworth Building, a towering 57-story affair that opened in 1913 on Broadway in lower Manhattan. Cass Gilbert's ornate design relied heavily on the Gothic cathedrals of Europe for inspiration; indeed, one onlooker referred to it as "the cathedral of commerce." Yet its ornate exterior belied perhaps the most advanced systems of any skyscraper to date, incorporating many of the technologies associated with more modern skyscrapers: a concrete caisson foundation, a braced frame to resist wind, and high-speed (local and express) elevators.

Two years later, in 1915, the Equitable Building—another downtown colossus—was completed on lower Broadway in New York. In contrast to the corporate headquarters

that had preceded it, the Equitable was a developer's building: it was designed to make money. It did that by maximizing the amount of office space, while keeping construction costs (and hence architectural details) to a minimum. The building rose 38 stories straight up from the street, in almost clifflike fashion, with little external adornment. A massive 1.2 million square feet (111,000 square meters) of office space lay within it in the form of an H, to allow light to penetrate beyond its soaring street walls.

The great mass of the Equitable Building, and the extent to which its towering walls blocked light and air, was not lost on the citizens or planners of New York City. It became the poster child for a burgeoning movement to adopt regulations governing the mass and height of the city's skyscrapers. One year after the Equitable opened, in 1916, the Building Zone Resolution—the first comprehensive zoning law in the country—was adopted in New York City.

The 1916 Zoning Resolution, as it is often called, did more than simply place limits on the design and height of tall buildings. To minimize the historic conflicts between residential, commercial, and industrial uses in various parts of the city, it assigned these uses to specific districts in each of the city's five boroughs. With respect to tall building construction, it stopped short of placing an overall cap on height (as Chicago continued to do) but instead permitted height to remain unrestricted over one quarter of the lot. It also established setback rules for specific tower heights, thereby ensuring that light and air would continue to reach the street and that towering street walls like the Equitable's could never be built again.

SINGER BUILDING
NEW YORK

METROPOLITAN LIFE INSURANCE TOWER
NEW YORK

WOOLWORTH BUILDING
NEW YORK

EQUITABLE BUILDING
NEW YORK

HALLIDIE BUILDING
SAN FRANCISCO

METERS

350 —
300 —
250 —
200 —
150 —
100 —
50 —

1910   1915   1920

New York's new zoning laws, and particularly its regulations governing setbacks, would have a major impact on the shape of buildings designed in the city from that point forward. Yet such was the demand for space that they did little to curb the growing number of tall buildings being constructed in the city. Throughout the 1920s commercial office towers reached record-setting heights—growing to 40 or 50 stories, albeit with slimmer, tapered towers and wedding cake-like, tiered façades.

Some rose much taller. The 71-story Bank of Manhattan Building opened on Wall Street at 927 feet (283 meters) in April of 1930, garnering the title of the world's tallest building. But it would hold that honor ever so briefly: one month later the 77-story Chrysler Building opened in Midtown. Competition between the two owners of these buildings was so fierce that the spire of the Chrysler Building was constructed in secret and then raised into place from within the building—to ensure that the tower would be able to claim the "world's tallest" title from its downtown competitor.

The extravagance of the commercial towers built in New York in the twenties as monuments to commerce was staggering. Some, like the American Radiator Building, contained dramatic display halls for corporate products—in this case heaters (when the building was sold to the American Standard Company, plumbing appliances replaced the heaters). Others commissioned promotional artwork that became an integral part of the building; the New York Telephone Company's building, completed in lower Manhattan in 1927, featured a lobby floor with bronze plates depicting the construction of the city's telephone network, as well as ceiling frescoes commemorating the history of communication.

Not all the towers that arose in New York and Chicago in the early part of the twentieth century were corporate trophies. Most were wholly speculative, and even the corporate towers contained sizable amounts of space that were rented out to others. Indeed, the tallest building constructed in New York during this period did not even carry a corporate name. The Empire State Building, constructed by a developer in just over a year, eclipsed both the Bank of Manhattan and Chrysler buildings in height when it opened in 1931. Its 102 stories rose 1,250 feet (380 meters) into the sky and contained over two million square feet (186,000 square meters) of office space—a vast amount even by today's standards. It would remain the tallest, and the largest, building in the world for almost four decades.

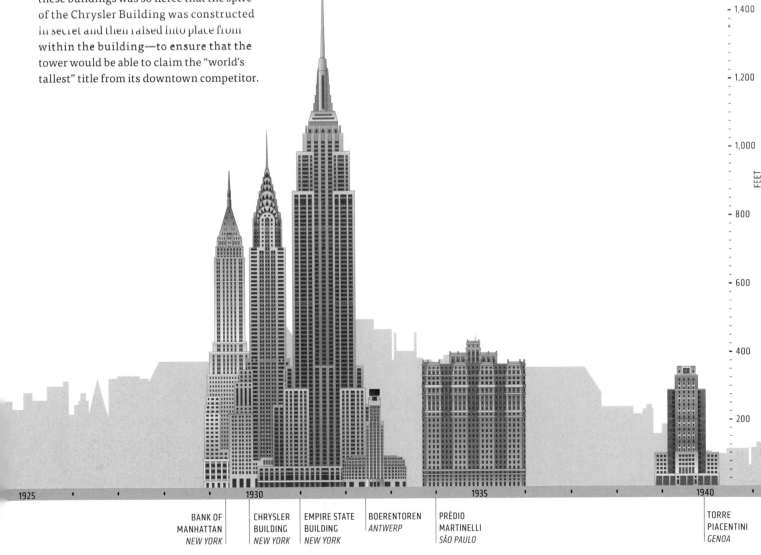

FEET

- 1,400
- 1,200
- 1,000
- 800
- 600
- 400
- 200

1925   1930   1935   1940

BANK OF MANHATTAN
NEW YORK

CHRYSLER BUILDING
NEW YORK

EMPIRE STATE BUILDING
NEW YORK

BOERENTOREN
ANTWERP

PRÉDIO MARTINELLI
SÃO PAULO

TORRE PIACENTINI
GENOA

The International Style

Throughout the first half of the twentieth century, skyscrapers remained almost exclusively an American affair. In 1930, 99 percent of the tallest 100 buildings in the world were in North America; the exception was the 28-story Prédio Martinelli, which opened in São Paulo, Brazil, in 1929. Europe remained generally resistant to the idea of constructing tall towers—in part for aesthetic reasons and in part due to concerns about safety. The earliest skyscrapers to appear on the continent were relatively modest affairs, such as the 26-story Boerentoren in Antwerp (1932) or the 31-story Torre Piacentini in Genoa (1940).

The Depression and Second World War put a halt to private-sector development around the world, and the race for height would not resume for decades. But the modern skyscraper did not stand still in the period following the war. In many ways, it changed more dramatically after the war than it had before it and in a primarily aesthetic sense—with a move away from the elaborate detailing of the early decades of the century to a much simpler, geometric form.

This modernist or international style approach to skyscraper design is generally attributed to Mies van der Rohe, a German architect who had proposed free-form towers encased only in glass panels as early as 1920. Mies's move to Chicago just before World War II set in motion a wave of European modernism perhaps best represented by his Seagram Building and by Skidmore, Owings & Merrill's Lever House, both completed along Park Avenue in Manhattan in the 1950s. These rather minimalist boxes, characterized by glass-sheathed steel and flat roofs, represented a dramatic break with the fussy art deco and gothic styles that had to that point characterized skyscraper design.

The advent of the glass "curtain wall," as the external skin of a building is defined, offered significant benefits to developers. Glass was less expensive than stone and allowed greater light penetration to the interior. Along with the development of fluorescent lighting and air-conditioning, the penetration of light permitted the design of larger blocks of contiguous, open space within commercial towers—which meant greater flexibility for users and more rentable area for developers.

While New York and Chicago were refining the aesthetics of the commercial tower, other parts of the world were just beginning to embrace the idea of height. The 37-story Torre de Madrid was completed in Spain in 1957, marking the earliest real skyscraper on the continent. By the mid-1960s, skyscrapers in the range of 30 to 40 stories were either being built or had been completed in other places as well—including Africa, the Middle East, and Australia.

If skyscraper aesthetics were changing, so too was the technology that underpinned their structure. In the late 1960s, a structural engineer named Fazlur Khan at Skidmore, Owings & Merrill in Chicago developed the idea of a "tubed" support structure—a building whose external perimeter walls would consist of a series of load-bearing tubes, instead of simple steel columns. These tubes could not only take greater weight, allowing buildings to reach higher, but because they could be arranged in a variety of shapes they also freed towers from the boxy, rectangular, or square shapes that had more or less defined the international style for two decades.

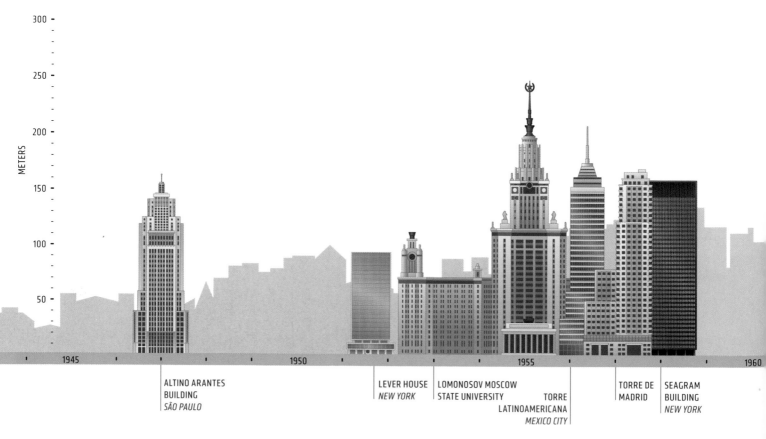

METERS

300
250
200
150
100
50

1945   1950   1955   1960

ALTINO ARANTES BUILDING
SÃO PAULO

LEVER HOUSE
NEW YORK

LOMONOSOV MOSCOW STATE UNIVERSITY

TORRE LATINOAMERICANA
MEXICO CITY

TORRE DE MADRID

SEAGRAM BUILDING
NEW YORK

The first tube-based skyscraper to open was, ironically, not one designed by Khan or his firm—it was the two towers of the World Trade Center in New York, which opened in lower Manhattan in 1972. At 110 stories and 1,365 feet (416 meters) in height, the buildings well exceeded the height of the Empire State Building and claimed the title of world's tallest buildings. But the Trade Center's claim would be short-lived. Two years later, in 1974, the Skidmore-designed Sears Tower opened in downtown Chicago. At 108 stories and 1,451 feet (442 meters), it became the world's tallest structure—and would hold on to that title for over two decades.

The openings of the World Trade Center and Sears Tower in the early 1970s marked a major departure from the signature headquarters buildings that had characterized the first 70 years of the twentieth century. The amount of square footage available in these towers was as extraordinary as their height: 8.6 million square feet (800,000 square meters) in the two towers of the World Trade Center and 4.5 million square feet (418,000 square meters) in the Sears Tower. The World Trade Center provided a commercial home to 430 companies and moved a staggering 50,000 people up and down in its complex system of shuttle and local elevators daily.

Supertall buildings (a phrase generally used to describe buildings over 1,000 feet or 300 meters in height) like the Trade Center and Sears Tower could be financed and filled only in a healthy real estate market. The oil crisis and ensuing recession of the mid-1970s halted the race for height, just as the Depression had in the 1930s. For roughly 20 years, no supertall skyscrapers

were planned anywhere in the world. When the market did begin to revive, in the 1990s, the United States was no longer the epicenter of activity. The new supertall towers were located in Asia and the Middle East, and they were no longer exclusively office buildings; most would be mixed-use ones, incorporating residential, retail, and commercial uses under one roof.

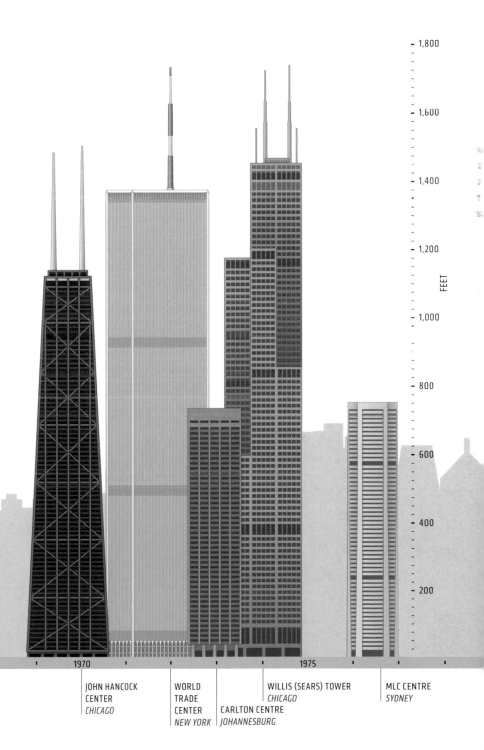

1965　　　　　1970　　　　　1975

JOHN HANCOCK
CENTER
CHICAGO

WORLD
TRADE
CENTER
NEW YORK

CARLTON CENTRE
JOHANNESBURG

WILLIS (SEARS) TOWER
CHICAGO

MLC CENTRE
SYDNEY

FEET

- 1,800
- 1,600
- 1,400
- 1,200
- 1,000
- 800
- 600
- 400
- 200

From Tall to Supertall

The geographic shift in skyscraper construction was dramatic and swift. In 1980, roughly 85 percent of the world's buildings over 500 feet (150 meters) were in North America; by 2008 that percentage had dropped to 28 percent.

The shift away from pure office buildings was equally marked. Whereas 85 percent of all skyscrapers in 1985 were used only as offices, by 2008 more than half of the world's skyscrapers incorporated residential and other mixed uses. These new towers were designed to function as veritable cities in the sky, hosting permanent residents as well as a host of retail features like health clubs, movie theaters, restaurants, and supermarkets. Multiple entrances to the buildings served to segregate users from one another.

The new skyscrapers were supertall, often rising over 1,000 feet (305 meters). New technologies, such as outriggers, brought added benefits: concrete cores attached to supercolumns on the perimeter, they allowed the new supertall buildings to be more open than tubed buildings,

typically characterized by a proliferation of smaller columns on the perimeter.

The leap in heights that the new technology permitted was significant. The Petronas Twin Towers, opened in Kuala Lumpur, Malaysia, in 1998, reached 1,483 feet (452 meters) in height to claim the title of world's tallest inhabited structure from the Sears Tower—though only 88 floors were occupied. Six years later, Taipei 101 opened in Taiwan at 101 stories and 1,670 feet (509 meters). It would hold the title of world's tallest for only five years. In early 2010, the 160-story Burj Khalifa opened in Dubai. A novel buttressed-core system allowed it to reach to 2,684 feet (818 meters), a leap of a full 1,000 feet (305 meters) over Taipei 101.

Many of the supertall towers built in Asia during the first decade of the twenty-first century could not claim the title of world's tallest, but they were enormous nevertheless. The Jin Mao Tower opened in Shanghai's Pudong district in 1988, at

88 stories. In 2008, the 101-story Shanghai World Financial Center opened across the street from Jin Mao. (A third tower, the 128-story Shanghai Tower, is under construction nearby.) And in Hong Kong, the 118-floor International Commerce Center opened in 2010—its height limited by law to that of the surrounding mountains.

These new towers, both in Asia and the Middle East, were remarkable for more than their height. In their architecture, many moved away from the simple geometric shape that characterized midcentury skyscraper architecture. Cultural references were deliberately incorporated into several of the new towers' designs. In Asia, these cultural references took the form of anything from pagodas (Jin Mao) to ancient Chinese symbols (Shanghai World Financial Center). In the Middle East, these cultural references often related to the geometric patterns typical of Islamic architecture.

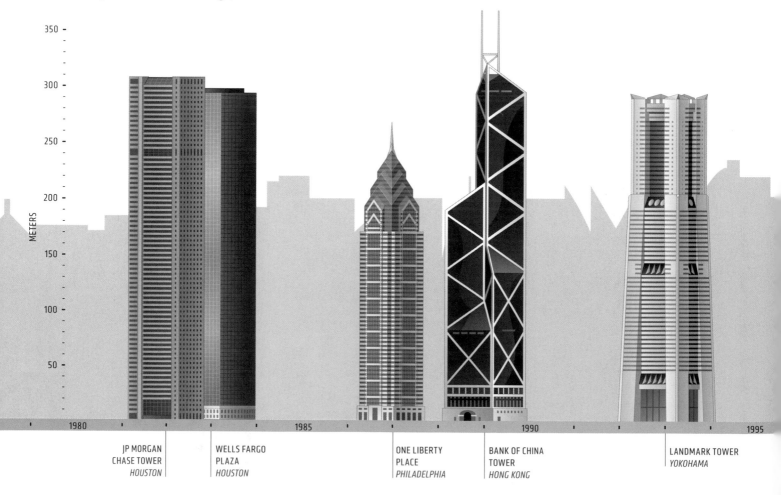

350					
300					
250					
200					
150					
100					
50					

METERS

1980   1985   1990   1995

JP MORGAN
CHASE TOWER
HOUSTON

WELLS FARGO
PLAZA
HOUSTON

ONE LIBERTY
PLACE
PHILADELPHIA

BANK OF CHINA
TOWER
HONG KONG

LANDMARK TOWER
YOKOHAMA

But notwithstanding their great heights and unusual shapes, today's skyscrapers are perhaps most remarkable for their complexity in operating around-the-clock as mixed-use destinations for thousands of people. The Burj Khalifa in Dubai, for example, contains 1,000 residences, 175 hotel rooms, and 37 floors of office space. As the centerpiece of the massive development known as Downtown Dubai, the complex at its base contains over 12 million square feet (1.1 million square meters) of stores, 22 movie theaters, a four-story fitness/recreation annex, and 3,000 underground parking spaces. To serve visitors and residents alike, the Burj offers over 27 acres (10.9 hectares) of park and the world's tallest observation deck.

Buildings like the Burj represent a new extension of the skyscraper as an urban form. Today's skyscrapers are not necessarily at the epicenter of a dense city, nor are they wholly devoted to office life—as they once were. Instead, mixed-use buildings like the Burj are increasingly embracing the residential aspects of urban life, as well as the commercial, recreational, and retail ones—becoming, in many ways, their very own "vertical cities."

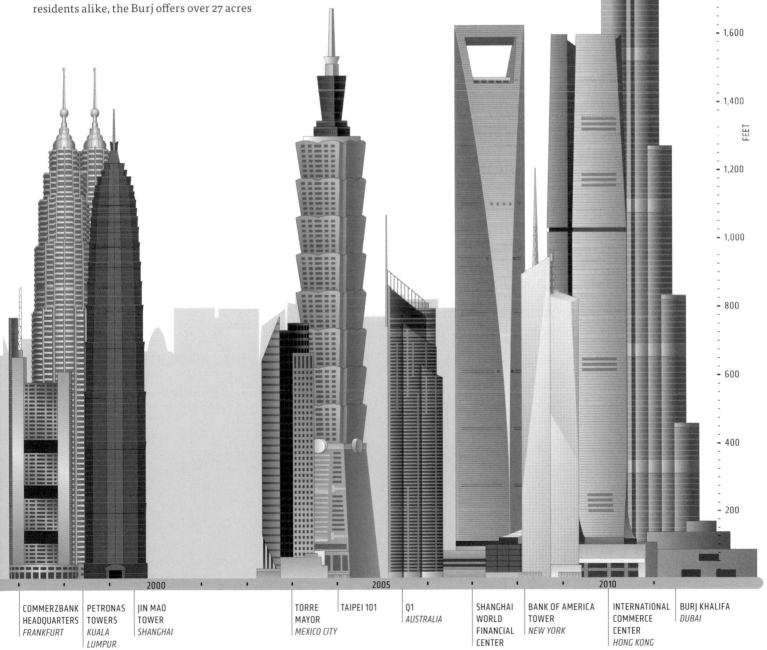

FEET

- 2,600
- 2,400
- 2,200
- 2,000
- 1,800
- 1,600
- 1,400
- 1,200
- 1,000
- 800
- 600
- 400
- 200

2000

COMMERZBANK HEADQUARTERS *FRANKFURT*

PETRONAS TOWERS *KUALA LUMPUR*

JIN MAO TOWER *SHANGHAI*

2005

TORRE MAYOR *MEXICO CITY*

TAIPEI 101

Q1 *AUSTRALIA*

SHANGHAI WORLD FINANCIAL CENTER

BANK OF AMERICA TOWER *NEW YORK*

2010

INTERNATIONAL COMMERCE CENTER *HONG KONG*

BURJ KHALIFA *DUBAI*

BUILDING IT

DESIGN

Skyscrapers, as one commentator noted, "are the ultimate architecture of capitalism." For over a century, architects at the top of their profession have been drawn to skyscraper design as they have to no other private commissions. Some, like Daniel Burnham and Cass Gilbert, made their names designing commercial towers. Others, like Mies van der Rohe, Richard Rogers, and Cesar Pelli, made their names on other types of buildings but have subsequently left significant marks on the skylines of the world's great cities.

The respect accorded these designers for their towers underscores the artistic aspects of designing skyscrapers, and some of the world's tallest towers can rightfully be considered a form of civic art. But skyscraper design is more about math and science than it is about aesthetics, due to the number of people these buildings house and the heights they must reach. It is also about money: skyscrapers are designed both to generate income and to serve as a long-term investment in urban areas where land values are high.

Skyscrapers are shaped by a series of very specific functional and financial considerations. The building must provide a certain amount of living or working space on a given site in a way that makes it attractive to tenants, at a cost that is affordable to the building's developer, and in a configuration that meets all building and planning regulations. It must also be structurally sound, withstand great gravity and wind loads, and operate seamlessly for thousands and thousands of users. While important in the marketing of a building, aesthetics are in many ways a secondary consideration.

The task of designing a building thus presents as much a managerial challenge as an architectural one, with each decision involving myriad experts from different fields. It may start with a vision, but that vision must be translated rapidly into specific answers to some very tricky questions: How tall should it be? How many floors should it have and how high should each one rise? How big should the floor plates be? What form should it take to best resist wind loads? How many elevators will be needed and where should they be located? What kind of skin should the building have—glass, brick, steel? How secure does it need to be?

Even the most talented architect could not answer these questions alone. While he or she takes responsibility for coordinating the design effort, it takes a sizable team to get it done: structural and mechanical engineers; elevator consultants, security and building code specialists, curtain wall and façade advisers, and office layout experts—to name but a few. Once these experts have done their work and an initial design is produced, construction estimators are hired to take a first look at how much that design will cost and how long it will take to build.

The process takes time—often multiple years. The first phase of the design process, often referred to as "schematic design," alone can take anywhere from a few months to a year. Developing the design further, known as "design development," takes many months more. It is very much an iterative process: designs that fail testing by the structural engineers must be reworked; designs that are too expensive must be tossed aside or rethought to bring them down in price; floor plans that don't pass muster with potential users must be rethought until they do. Thousands of hours of consultant time and millions of dollars of the developer's money will be spent before the first shovel touches the ground.

Poets in Steel

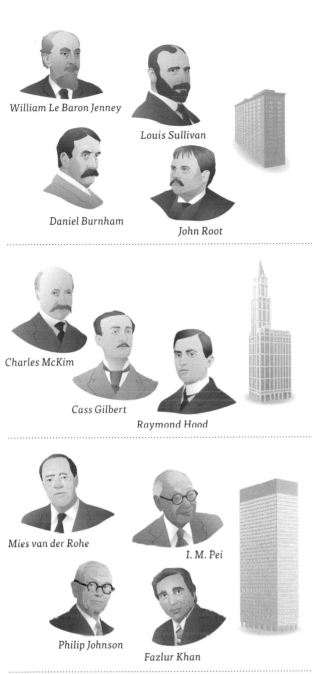

William Le Baron Jenney

Louis Sullivan

Daniel Burnham

John Root

Charles McKim

Cass Gilbert

Raymond Hood

Mies van der Rohe

I. M. Pei

Philip Johnson

Fazlur Khan

William Pedersen

Norman Foster

Cesar Pelli

Rafael Pelli

1880
1900
1920
1940
1960
1980
2000

■ FIRST CHICAGO SCHOOL

Also known as the "commercial style," the First Chicago School grew out of the structural innovations pioneered by William Le Baron Jenney. The style was further developed by Jenney's protégés Louis Sullivan and Daniel Burnham—both of whom performed their architectural internships at his firm—as well as John Root. The Home Insurance Building of Chicago was designed by Jenney and exhibited many of the distinguishing features of this style, including a steel frame with simple masonry cladding and large plate-glass windows.

■ BEAUX-ARTS

"Beaux-Arts" refers to the neoclassical architectural style taught at the École des Beaux-Arts in Paris in the late nineteenth century, which heavily influenced U.S. architecture between 1880 and 1920. The best known architectural firm specializing in the Beaux-Arts style was McKim, Mead, and White; Cass Gilbert spent his early career there before starting his own practice and subsequently designing the neogothic-inspired Woolworth Building. Raymond Hood's art deco masterpieces also characterized the period. More ornamental than the practical Chicago School style, Beaux-Arts buildings featured arched windows and doors, classical architectural details, and statuary and sculpture.

■ INTERNATIONAL STYLE AND SECOND CHICAGO SCHOOL

The international, or modernist, style, also known as the Second Chicago School, emerged initially in the 1920s and 1930s; however, the Great Depression and World War II prevented it from flourishing. The 1950s completion of the United Nations headquarters, the Seagram Building, and Lever Building in New York—all distinguished by their boxy shapes and sleek glass façades—ushered in this new era. Best known among the internationalists were Mies van der Rohe, I. M. Pei, and Philip Johnson. Around the same time, Fazlur Khan, a structural engineer working for Skidmore, Owings & Merrill in Chicago, developed the concept of framed and bundled tubes—which would allow the skyscrapers of the late sixties and early seventies to reach even higher into the sky.

■ SUPERTALLS AND GREEN

At the end of the twentieth century, the quest to build taller expanded worldwide. Drawing on the structural innovations developed by Khan, Cesar Pelli designed the record-setting Petronas Towers in Malaysia. A decade later, Bill Pedersen and his colleagues at Kohn Pedersen Fox designed the supertall Shanghai World Financial Center in China. Simultaneously, as an awareness of global warming gave rise to the "green building" movement, architects began to look for ways to make buildings more energy efficient. The green skyscraper designs of Sir Norman Foster, like those of Cesar Pelli's son, Rafael, ushered in a new era of sustainable building design.

Selecting a Design

Even on a heavily constrained site in the middle of a dense city, a skyscraper can take many forms to meet a developer's programmatic needs. Historically, skyscrapers in Chicago and New York have either taken the form of a tower on a podium or have risen straight up from the street. They typically have been square or rectangular, although more recently some have taken a more elliptical shape. Most frequently, they have ended in an elegant spire or been crowned by a blunter, functional top.

Today there are many, many more options for skyscraper forms. Thanks to technological advances, towers can bend back on themselves or incorporate large openings to decrease wind resistance. Their service cores may be on the outside of the structure rather than in the middle of it. They can twist and turn, and—in seeming defiance of the laws of gravity and conventional building wisdom—can even be larger at the top than they are at the bottom.

In general, residential buildings tend to be slimmer than commercial ones. Residential floor plates are smaller, so as to allow natural light into most, if not all, rooms. In commercial buildings, natural light is less of a consideration and large, flexible floor plates are more likely to meet tenants' needs.

But there are few other rules that define what shape tower is appropriate for a particular site. So long as the developer's functional and financial needs are met, the choice of form is largely an aesthetic one. In some cases, a particular architect will be chosen by a developer and asked to provide one or more designs. In other cases, developers will hold an informal competition among short-listed architects to select a preferred design—sometimes paying them a stipend for their time and effort.

A tower for the bus terminal

In 2008, the Port Authority of New York and New Jersey and its partner, Vornado Realty Trust, invited four firms (a fifth was added subsequently) to submit preliminary designs for a tower of roughly 1.6 million gross square feet (150,000 square meters) that would sit atop a newly renovated wing of the Port Authority Bus Terminal at West 42nd St. and 8th Ave. in Midtown Manhattan. This north wing of the bus terminal had been constructed in the 1970s with the knowledge that a tower would someday rise above it; hence, the structural supports to take the gravity load were built into the terminal. But even with the limitations of an already built foundation, the designs submitted by some of the world's foremost commercial architects differed radically from one another.

PELLI CLARKE PELLI (PCP)
This elliptical-shaped tower design presented a dramatic contrast to the rectilinear bus terminal podium upon which it would stand. The skin of the office building appeared to wrap around the structure in a series of folded leaves projecting upward from the base. The lobby of the building would be located on West 42nd St., a main thoroughfare that offers easy access to crosstown transportation.

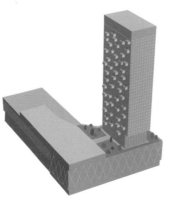

SHoP ARCHITECTS
The design for this tower was the "greenest" of the four entries. The firm designed a classic rectilinear tower but incorporated a system of hanging gardens on its southern face. A sky lobby was located on the fifth floor of the terminal, where a car drop-off facility would operate in conjunction with the terminal's existing ramp structure. Double-paned glass was specified to allow for better management of solar and thermal impacts.

TALLER DE ENRIQUE NORTEN ARQUITECTOS (TEN ARQUITECTOS)

A building that literally bent in the middle— not once, but twice—was featured in this submission. External bracing needed to stabilize the structure featured prominently in the building's appearance. The building's lobby would be located on the corner of 8th Ave. and West 41st St., the latter an underutilized street that offered capacity for taxi drop-off and black-car queuing. A publicly accessible sky terrace projected south from the building's fifth floor across the roof of the southern half of the bus terminal.

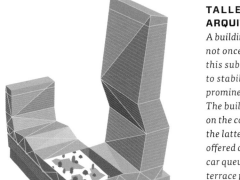

N

KOHN PEDERSON FOX (KPF)

This tower design was composed of two vertical elements: one sized to reflect the old McGraw-Hill Building adjacent to the tower's western flank along West 42nd St. and one that reached higher, to reference Renzo Piano's New York Times Building across the street on 8th Ave. They rose as part of a series of cubes of different sizes, mimicking the cubist form of the podium at the building's base. The initial design included a series of atria midway up the tower, to be used by corporate tenants as points of connectivity between their office staff on separate floors.

ROGERS STIRK HARBOUR + PARTNERS

This entry was ultimately awarded the schematic design contract—after each of the four original designs was ruled out on account of cost, aesthetic, or constructability issues. The design produced by the Rogers team resembled none of the other proposals. To ensure that the flow through the bus terminal below would not be impeded by penetrations from above, the firm chose a unique side-core design that featured elevator banks on the north and south extremities of the building. This resulted in large, unobstructed floor plates that would provide flexibility to potential office tower tenants looking to maximize internal layout options. To segregate office workers physically and psychologically from the busy bus terminal, a sky lobby—accessed through a bank of shuttle elevators at the building's 8th Ave. street-level entrance—was designed to welcome workers on the fifth floor of the structure.

Designing the Shanghai World Financial Center

The construction of the Shanghai World Financial Center, the world's third-highest building, highlights just how complex skyscraper construction—at the intersection of engineering, architecture, economics, and even politics—can be. Begun in the early nineties, when Japanese developer Minoru Mori hired Kohn Pederson Fox to design a building in Shanghai's Pudong district, the building was not completed until 15 years later.

INSPIRATION

Inspiration for a tower's form can come from many places. In the case of the Shanghai World Financial Center, architect Bill Pedersen's inspiration came from ancient China, when the earth was represented by a rectangular prism and the sky by a circular disk.

Two gently sweeping arcs rise in prismlike fashion from the ground to converge near the top of the tower, representing the confluence of the earth and the sky.

STRUCTURE

The building's foundation was laid in 1997, but the Asian economic crises brought a halt to the project. When it resumed, in 1999, the developer wanted a bigger building—a 15 percent increase in gross area. Building bigger on the existing foundation required decreasing the weight of the building by 10 percent and redistributing its load.

So a new structural system was devised: by introducing outrigger trusses—which connect the building's core with megacolumns on its corners—the size of the core walls and the weight of the steel on the perimeter were significantly reduced.

Design Development

The choice of a design that meets a developer's programmatic needs, in terms of the amount of square footage on each floor and the type of use, and shares the company's aesthetic sensibility is an important first step in building a skyscraper. But the selection of a conceptual design represents only the beginning of a design development process that typically takes anywhere from 12 to 18 months. During that period the design will be refined many times and in many ways—in response to the building's structural needs, to its projected cost, or simply to the changing aesthetic whims of either the developer or potential tenants.

Sometimes the design process takes even longer, because design criteria change along the way or because the project is put on hold for some reason. Many of the building projects halted in the wake of the 2008 recession may take a different form than originally envisioned when they are ultimately constructed.

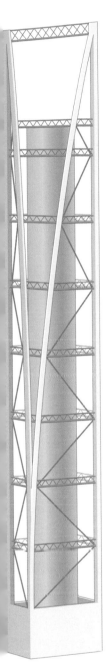

WIND LOAD

The structural system was not the only component of the building that would witness significant change during the design process. To reduce lateral loads on the building as much as possible, a circular hole (referencing a traditional Chinese moon gate) was placed atop the building.

It allowed the wind at higher elevations to pass through, thereby reducing the lateral pressure it exerted on the structure.

POLITICS

However, the circular hole came under criticism for evoking the rising sun of the Japanese flag—a sensitivity no doubt related to the nationality of the developer and the history of Sino-Japanese relations.

At first, ideas of reducing its circularity by introducing a promenade across the circle, or a Ferris wheel within it, were explored. These were ultimately dismissed in favor of a trapezoidal hole.

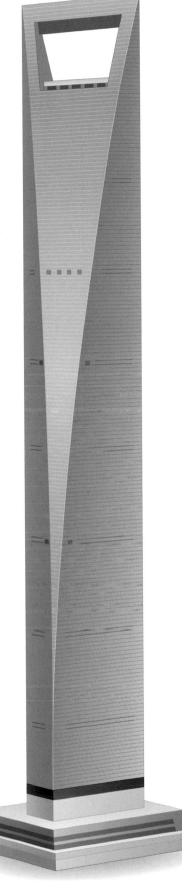

Zoning

The visions of skyscraper architects are colored by a variety of things, many of them engineering-based, such as the weight-bearing capacity of foundations or the flexibility of certain materials. But they are also shaped by rules and regulations governing land use—including building codes, height restrictions, and what is commonly known in the United States as zoning.

Zoning and land-use regulation are important in understanding skyscraper design for several reasons. They determine the use allowed on a particular site (for example, residential or commercial), the size of a building, and sometimes even its shape.

The world's first zoning ordinance was passed in New York in 1916, on the back of an outcry over the Equitable Building's towering façade and the seven-acre shadow it cast over lower Manhattan. Since that time, zoning has been a key land-use tool for urban planners around the world, both to protect against incompatible uses and to ensure that light and air continue to reach city streets in even the densest of urban areas.

Modern zoning and land-use regulations are more refined tools than they were a century ago. They still protect against incompatible uses, but they may also incorporate parking requirements, sustainability practices, street-wall heights, sidewalk widths, and even rules relating to advertising on building façades.

In the last half century, zoning has also become a tool to incentivize development of a certain type of building or of related amenities, such as parks or public plazas. Today it is actively used to encourage the mix of residential and commercial uses now thought of as central to successful neighborhoods as well as to facilitate special forms of development at geographically unique locations, such as transit hubs or the waterfront.

A sample of Manhattan zoning

In Manhattan, a variety of district designations govern the size and use of buildings. Designations beginning with an "M" signify a manufacturing district; those beginning with an "R" signify residential; and those beginning with a "C" signify a commercial district. Generally, the letter designation will be followed by a number that conveys the intensity of activity permitted in that district: R8 districts, for example, permit midrise (eight- to 10-story) buildings, while R10 districts allow higher floor-area ratios (FAR). The FAR determines the amount of floor area permitted as a multiple of the area of the building lot.

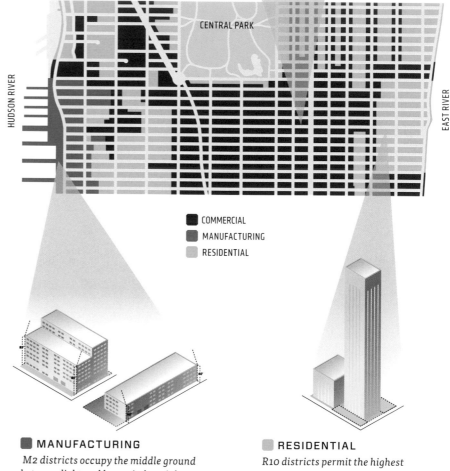

CENTRAL PARK

HUDSON RIVER

EAST RIVER

■ COMMERCIAL
■ MANUFACTURING
■ RESIDENTIAL

■ **COMMERCIAL**
Most C6 districts are areas well served by mass transit. Corporate headquarters, large hotels, entertainment facilities, retail stores, and high-rise residences in mixed buildings are permitted a maximum floor-area ratio (FAR) of 10 or 15 times the size of the base site.

■ **MANUFACTURING**
M2 districts occupy the middle ground between light and heavy industrial areas. They are mapped mainly in the city's older industrial areas along the waterfront, including Manhattan's Hudson River piers, and have varying floor-area limitations.

■ **RESIDENTIAL**
R10 districts permit the highest residential density in the city—a FAR of 10 that can be increased to 12 if provisions for affordable housing are included. In Manhattan, much of Midtown and downtown, as well as major crosstown streets and avenues, permit R10 density.

Incentives

The phrase "incentive zoning" is often used to refer to floor-area bonuses that are given to developers in exchange for the provision of certain public amenities. Perhaps most common among the amenities that planners aim to provide is affordable housing. Developers of residential towers will commonly be incentivized (and sometimes required) to build affordable housing at or near the site of their tower or, alternatively, pay a certain amount into a civic housing fund.

Other amenities that have commonly been incorporated as bonus items in zoning codes include public plazas and space, parking, transit improvements, and sustainability goals. In return for providing one or more pubic amenities, developers are generally given the right to build out additional floor area (i.e., the FAR is raised on that particular site).

CHICAGO: PUBLIC SPACE

In Chicago, developers who provide public plazas or arcades have been eligible for floor-area bonuses since the 1940s. Current zoning rules encourage the provision of additional amenities including green roofs, underground parking, and water features.

NEW YORK: SUBWAY IMPROVEMENTS

In New York, developers who provide a series of major improvements to the subway network at or near their site may take advantage of a bonus of up to 20 percent of the base floor area otherwise attributable to their site.

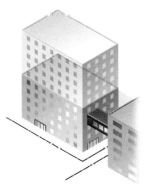

HONG KONG: CONNECTIVITY

In Hong Kong, developers who provide public plazas or elevated walkway connections between office buildings in the downtown area are eligible for a floor-area bonus.

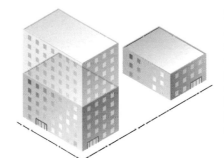

DENVER: HOUSING

In Denver, any developer of rental housing who provides affordable units receives a 10 percent density bonus, a subsidy toward the affordable housing, a reduction in parking requirements, and an expedited permitting process.

Building tall in London

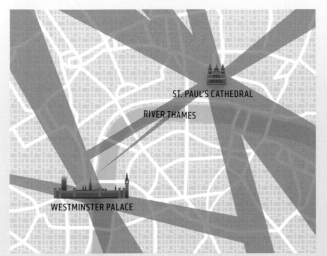

ST. PAUL'S CATHEDRAL

RIVER THAMES

WESTMINSTER PALACE

In certain cities around the world, the criteria for locating skyscrapers are less quantitative and more qualitative than the FAR-based zoning calculations common in parts of the United States. In London, for example, new towers need to meet certain subjective criteria: they must prove attractive as landmarks, prompt local economic development, and be sensitive to the character of their surroundings.

The city is particularly concerned about the impact new development might have on existing view corridors—particularly views of St. Paul's Cathedral or the Palace of Westminster. Buildings that affect these views, either by blocking the landmark itself or by projecting into either the foreground or the background of a particular view, are scrutinized carefully. The impact on other "protected vistas," including London panoramas, river prospects, townscape, and linear views, is also taken into account during the approval process.

■ VIEW CORRIDORS

The Core

The design of almost any high-rise structure starts with locating the core—the area of a skyscraper that houses elevators, fire stairs, bathrooms, machine rooms and utility closets. "Risers," which bring power, water, air, and telecom to each floor, are also located within the core.

The configuration and location of the core will vary from building to building, depending on its use and size. In the United States, cores are generally located at the center of commercial buildings, but they may also be placed to one side (a side-core building) or in multiple locations (more common in residential towers). Wherever it is, the core has a major impact on how each floor of a skyscraper is laid out.

The core does more than just provide a home to the building's pipes and elevators. It can also serve as the main structural element allowing a skyscraper to withstand both horizontal and lateral forces. Generally made of a combination of steel and concrete, the cores of most buildings play a critical role in providing the stiffness needed to withstand the force of strong wind gusts—particularly for higher floors.

Designing a core is very much a mathematical exercise and involves coming up with a plan for a typical floor (a "floor plate") that meets a developer's programmatic needs. Two critical features in determining the floor plate in commercial towers are lease depths and the number of elevators required to service the building's projected population.

Lease depths, the amount of usable space from the exterior wall to the edge of the core, can range anywhere from 20 to 50 feet (6 to 15 meters). In commercial buildings in the United States, lease depths typically fall between 40 and 50 feet (12 to 15 meters). European floor plates tend to be shorter to allow more penetration of natural light, while commercial buildings in Asia often have deeper floors.

Elevators are equally important to core design and floor layout. The number and size of passenger elevators can vary dramatically based on building type: hotel, residential, and office towers have very different patterns of peak elevator usage and demand. Freight elevators, as well as firefighter lifts (used in certain countries by fire personnel in an emergency), vary in size and number as well and must also form part of the core.

In addition, considerations must be given to handicapped accessibility. In the United States, the Americans with Disabilities Act (ADA) sets specific accessibility guidelines for public spaces and commercial facilities. This act includes requirements on the widths of doors, the number and types of elevators, wheelchair-accessible toilet room fixtures, the design of fire stairs, and the nature of floor surfaces.

Anatomy of a core

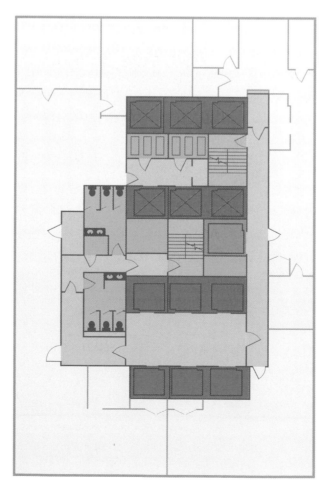

■ EGRESS

Like other buildings, skyscrapers must have an unobstructed way for people to exit the building in an emergency. If more than one exit access is required, two exits must be separated by a minimum distance that is based on the dimensions of the floor.

■ FREIGHT ELEVATOR
■ PASSENGER ELEVATOR
⊠ ELEVATOR NOT SERVING FLOOR

Elevator shafts and lobbies consume most of the space in a building's core and will typically feature two or more banks of cars servicing different floors. Inactive elevator lobbies are often used for toilet rooms.

■ RISERS

Also known as service cores, these vertical penetrations allow electrical cables, plumbing, heating, cooling and ventilation ducts, and piping to find their way from the street to dedicated mechanical floors or rooms and occupied floors.

■ TOILET ROOMS

While "potty parity" laws dictate the equal or equitable provision of toilets for women and men within public and commercial spaces, the American Rest Room Association recommends a greater number of toilets for women as the number of people on a given floor grows.

# OF PEOPLE	# OF WOMEN'S & MEN'S TOILETS RECOMMENDED	
400–500	🚽🚽🚽 🚽🚽🚽	🚽🚽 🚽🚽
300–400	🚽🚽🚽 🚽🚽	🚽🚽 🚽🚽
200–300	🚽🚽 🚽🚽	🚽🚽 🚽
100–200	🚽🚽 🚽	🚽🚽 🚽
50–100	🚽🚽	🚽🚽
0–50	🚽	🚽

Building typologies

Cultural factors have a significant impact on the layout and design of skyscraper floors, and hence on a building's ultimate appearance. As a result, significant differences in the typical size and shape of buildings can be found around the world.

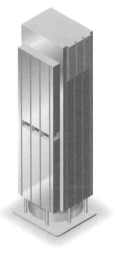

 FLOOR PLATE
 CORE

UNITED STATES

In the United States, a tradition of relatively short commercial leases (often 15 years) results in emphasis being placed on the flexibility of floor plate design. Mechanical ventilation is expected to be provided, as is redundancy of power and telecom systems, so the amount of space given over to mechanical space on each floor can be significant.

EUROPE

Continental Europe still relies heavily on natural ventilation rather than air-conditioning—in part because of the milder summers. Office workers in the European Community have a right to work within 25 feet (eight meters) of natural light, so floor plates tend to be smaller and towers more slender. This shape is conducive to private offices, which are more prevalent than they are in the United States.

ASIA

In Japan, skyscrapers are often more bulky than those in the United States. This is largely due to the value placed on structural strength and the assumption that wider buildings will perform better in earthquakes than thinner ones. Contractors play a larger role in determining the design of a building than they do in the United States or Europe, further contributing to this emphasis on structural integrity.

Test fit

Most commercial spaces are offered by landlords as "core and shell," with few finishes to the space other than common areas. Prior to renting a space, most commercial tenants will hire an architect to develop a "test fit" to determine whether the space suits their operational needs.

The way tenants use a space can vary dramatically. Law firms typically tend to favor exterior offices, whereas many media and design firms favor open, collaborative spaces. Financial service firms may require large, open trading floors to accommodate several hundred traders.

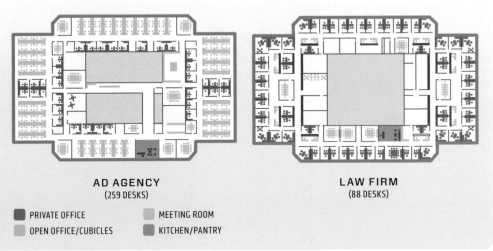

AD AGENCY
(259 DESKS)

LAW FIRM
(88 DESKS)

■ PRIVATE OFFICE ■ MEETING ROOM
■ OPEN OFFICE/CUBICLES ■ KITCHEN/PANTRY

Stacking

Gone are the days when companies could afford to build skyscrapers simply to market their image or brand. Today, with very few exceptions, skyscrapers are built to make money. A corporate headquarters building must prove itself cheaper than the cost of renting space elsewhere—and those that do will typically not be built by the corporation itself but by developers with expertise in managing the complex process of development.

Before any skyscraper is built, a developer's architect will complete a stacking diagram, which identifies the amount of floor area on each floor and its intended use. Typically, underground areas will be used for parking, storage, and mechanical support. One or two levels at or just above the street will be used for retail, which tends to generate greater revenues than any other ground-level use. The remainder of the tower may be devoted solely to commercial offices or developed as a hotel or residential property, with units to be sold or rented. In some cases, skyscrapers feature both commercial and residential or hotel uses stacked on top of one another—in what has come to be known as a "mixed-use building."

The amount of space devoted to mechanical infrastructure ranges widely from building to building. The number and location of mechanical floors is generally determined by the size of the building and the amount and type of services demanded by tenants. It may also be a function of the sophistication of municipal infrastructure, and the need for backup power generation or water supply.

The stacking diagram incorporates all of these uses and is an important tool in calculating the profitability of a prospective building. It assigns specific amounts of square footage to both revenue- and nonrevenue-producing uses, each of which carries with it a specific build-out cost and a specific forecast for future revenue. The total square footage assigned to a use, together with the per-square-foot (psf) costs and revenues associated with that use, form the basis for a building pro forma that represents the overall economics of the project.

FLOOR	FLOOR AREA (SQ. FT.)
40	25,000
39	25,000
38	25,000
37	25,000
36	25,000
35	25,000
34	25,000
33	25,000
32	25,000
31	
30	35,000
29	35,000
28	35,000
27	35,000
26	35,000
25	35,000
24	35,000
23	35,000
22	35,000
21	35,000
20	35,000
19	35,000
18	35,000
17	35,000
16	35,000
15	35,000
14	35,000
13	35,000
12	35,000
11	35,000
10	35,000
9	35,000
8	35,000
7	35,000
6	35,000
5	35,000
4	
3	55,000
2	55,000
1	55,000

Stacking it up

Most skyscrapers are designed to incorporate multiple types of uses, generally stacked one above the other. Historically, towers included retail below and either residential or commercial space above. Today, an increasing number of tall buildings incorporate both uses above a retail base.

▪ HOTEL / RESIDENTIAL

In a mixed-use skyscraper, hotel or residential uses are typically placed at the top of the tower. Placing these uses higher in the building allows the developer to command better rental rates or sales prices by selling the superior views and prestige of being "on top." Practically, residential and hotel users place less demands on an elevator system and will therefore consume less space on the lower floors for shafts.

▪ MECHANICAL

The location of mechanical floors will be a trade-off between proximity to infrastructure at the street level and proximity to end users on the occupied tower floors.

▪ OFFICE

Office uses place the heaviest demand on elevator systems due to the higher density of workers and peak use demands in the morning, at noon, and in the evening. Intensive uses like trading floors will be located lowest in the building and may even require dedicated elevator systems.

▪ RETAIL

In most cities, street-level retail will command the highest rents on a per-square-foot basis. However, rents drop off significantly as distances from the street increase, so retail will typically only be located on the first two to three floors of a skyscraper podium.

What would it look like in the suburbs?

A notional 1.3 million square foot mixed-use skyscraper with a FAR of 15 will cover 60 percent of a typical New York City block. The same 135,000 square feet of retail (12,500 square meters), 225,000 square feet of residential (21,000 square meters), and 875,000 square feet of office (81,000 square meters) spread over a typical suburban setting with strip malls, one-quarter-acre (0.1 hectare) building lots, and open parking would cover more than 21 New York City blocks.

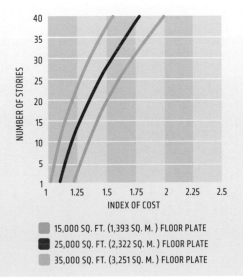

How high should it be?

The decision as to how tall a skyscraper will be is only rarely an aesthetic one. Developers first determine the "economic height" of a building: the number of stories that will produce the greatest return on the money invested in the building's construction and operation.

As buildings grow taller, they demand stronger foundations and bracing, larger mechanical systems, and—of primary importance in the days of supertall buildings—more elevators. Quite apart from the costs associated with all of this extra construction, each additional bank of elevators reduces the floor area available

for rent or sale—thus pushing down the revenue potential for the building.

The ideal height for a skyscraper will vary from place to place, depending on the cost paid for land, the size of the plot, the level of rents, zoning restrictions, and the cost of construction. In assessing the economic virtues of one design against another, a developer will usually assess the cost per foot (both the "hard costs," including materials and labor involved in construction, and the "soft costs," including financing, permitting, and design work) and compare it to the revenue per foot to be generated in rent or sales.

NUMBER OF STORIES (y-axis): 1, 5, 10, 15, 20, 25, 30, 35, 40
INDEX OF COST (x-axis): 1, 1.25, 1.5, 1.75, 2, 2.25, 2.5

■ 15,000 SQ. FT. (1,393 SQ. M.) FLOOR PLATE
■ 25,000 SQ. FT. (2,322 SQ. M.) FLOOR PLATE
■ 35,000 SQ. FT. (3,251 SQ. M.) FLOOR PLATE

Making It Memorable

Even with all the constraints placed upon them, such as zoning rules, building codes, structural requirements, and above all the need to make money, architects have—for more than a century—found ways to realize their visions in the sky. In the early decades of the twentieth century it was the building's crown that proved most distinctive: the spires of the Woolworth and the Chrysler buildings, for example, speak for themselves. Today, thanks to advances in electronics, materials science, and construction, there are many different techniques employed to make a skyscraper stand out from its peers—including the way it is lit, the signage that is placed at its base or top, and the nature of its silhouette against the skyline.

Lighting

EMPIRE STATE BUILDING

The lights at the top of the Empire State Building change almost daily to celebrate holidays (e.g., green for St. Patrick's Day), sports events (e.g., blue and white for a Yankees championship), or official visitors (e.g., blue for the United Nations General Assembly).

BURJ AL ARAB

This Dubai hotel is shaped like the sail of a dhow, with two "wings" spread in a V to form a vast "mast." At night, a complex arrangement of projected lighting makes the Burj a changing beacon seen from outside and provides a dramatic show for those inside its atrium.

COMMERZBANK HEADQUARTERS

The exterior lighting of this Frankurt tower was designed by famed lighting artist Thomas Emde, with changing hues of gold that complement and highlight the overall design of the building.

Signage

Sometimes buildings are notable simply because of the signs placed upon them. Certainly this is true of New York's Times Square, where skyscrapers compete for the most innovative signage, ranging from advertising to news tickers to sports clips. But skyscraper signage may also be found at the top of the building, notable only from a distance or from an airplane. In most cases, the right to stick a corporate name atop a building comes with ownership—or at least a long-term lease.

New York

London

Crowns

TRIUMPH PALACE

The tallest tower in Europe when completed in 2006, this Moscow tower is sometimes called "Stalin's Eighth Sister" after the similarly designed Seven Sisters built during the Stalin era.

WOOLWORTH BUILDING

The neogothic New York landmark is topped by a copper pyramidal cap and distinguished by overscaled finials and crockets that add to its upward movement.

CHRYSLER BUILDING

The distinctive spire of this art deco New York icon was secretly built inside of the building and raised at the last minute to win a crosstown height race with the Bank of Manhattan Building.

TORRE CAJA MADRID

This Madrid headquarters building is essentially a tall arch, with the service and circulation cores framing uninterrupted office floors at its center.

CITIGROUP CENTER

The slanted roof of this New York skyscraper, designed originally to house solar panels, was a departure from the prevailing flat-roof international-style buildings of its time.

TRANSAMERICA PYRAMID

The shape of this San Francisco landmark allows more air and light to reach the street. Two wings on either side accommodate elevators and egress stairs.

Pairs

EMIRATES TOWERS

While the Jumeriah Emirates Towers Hotel in Dubai has more floors, its companion Emirates Office Tower is taller due to larger floor-to-floor heights.

BAHRAIN WORLD TRADE CENTER

The sail-shaped towers of the Bahrain World Trade Center are connected via three sky bridges, each holding a wind turbine.

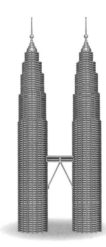

PETRONAS TOWERS

The Petronas Towers in Malaysia feature a sky bridge that links the forty-first and forty-second floors and can be slid in and out of the towers to prevent breakage during high winds.

FOUNDATIONS

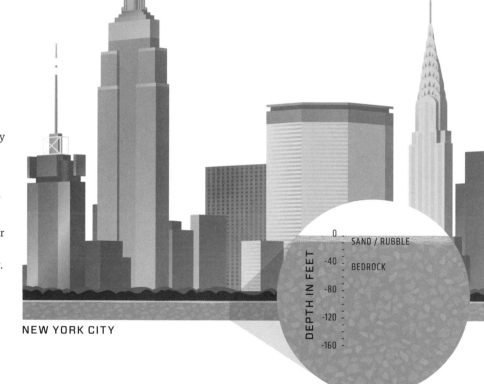

NEW YORK CITY

DEPTH IN FEET

0 — SAND / RUBBLE
-40 — BEDROCK
-80
-120
-160

 **N**othing of any size will remain standing for long without a foundation—from a hole in the sand dug to support a beach umbrella to the metal supports for a highway sign. Foundations do many things: they keep water out, support gravity loads without settling unduly, resist the "uplift" forces of wind, and hopefully stabilize a structure for a very long time. Even the Leaning Tower of Pisa's foundation, while clearly not perfect, has helped keep it from falling over.

Foundations have been important to city building for centuries. Most of the world's earliest metropolises were ports, located adjacent to lakes, rivers, or the sea, and as a result their soil tended to be watery and unstable. Wood piles were used as foundations in these places until the Industrial Revolution, when iron became more widely available.

The earliest skyscraper foundations were thus made of iron, soon to be supplanted by steel. At the time the science of geotechnical engineering was in its infancy. Perhaps not surprisingly, the earliest tall buildings—particularly a number of those built on the soft clay soil of Chicago—settled more than they should have. The Federal Building in Chicago settled so much after its completion in 1880 that it had to be completely destroyed 18 years later.

But foundation technology moved forward quickly as buildings grew taller.

The size and shape of larger loads required foundations to reach farther down into the soil, typically penetrating multiple layers of the earth. In general, the taller a building the deeper its foundation would reach—and the more complex the foundation design and construction process that accompanied it.

Today the amount a skyscraper will settle is carefully calculated and factored into the design of the building. It is only one aspect of a very complex design process that demands myriad decisions, including what kinds of supports should be used underground (e.g.,

piles or caissons), what types of material they should be made of (e.g., steel or concrete), how deep they should extend (e.g., to bedrock or not), and how they should be placed underground (driven or bored).

In general, this process is highly scientific, with powerful computers used to model and predict a building's behavior. However, there is an element of art to it too: it is said that the world's best geotechnical engineers can estimate the extent of building settlement fairly accurately on the back of an envelope, given the basic

Test boring

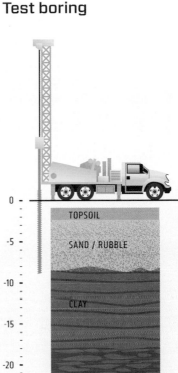

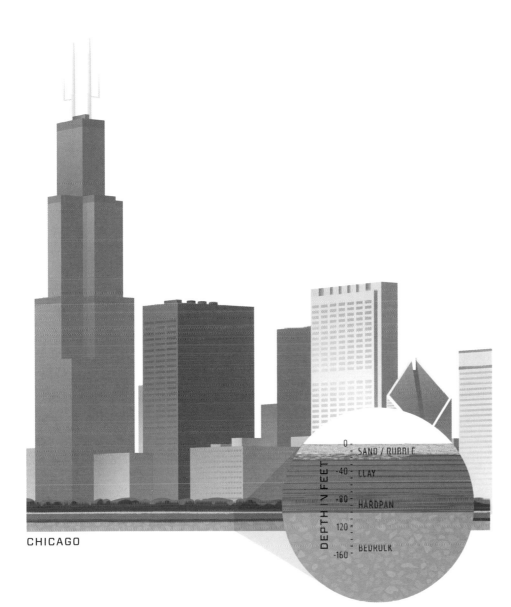

CHICAGO

parameters of the soil and the building's configuration and loads.

Soil types and layers vary greatly from city to city. Within cities themselves, soil layers and levels can vary dramatically; in New York, for example, bedrock is much closer to the surface uptown, near Central Park, than it is in lower Manhattan. And even within the confines of a building site, soil irregularities can require modifications to standard designs; Malaysia's two Petronas towers were moved 200 feet (60 meters) to take account of an unusually

precipitous drop in the bedrock at the point where one set of piles was to be located.

Despite the urban myth that skyscrapers can only be built where bedrock is close to the surface, they can be built just about anywhere. Two of the world's tallest stand proudly near Chicago's Lake Michigan—on soft, claylike soil. Two more are founded on the sandy soil of what was once Lake Texcoco in Mexico City and have proven themselves (not once, but three times) to be stronger and more stable than many of their shorter neighbors.

To design a foundation, geotechnical engineers will prepare a detailed map of the soil beneath a building site. They are particularly interested in the location of different layers of rock or minerals and the extent to which air and water may be present. Several techniques may be used to prepare this map, including seismic measurement, ground-penetrating radar, and test borings.

Test borings involve forcing a pipe casing into the earth to remove samples of material across the building site. Samples will be tested for weight, porosity (how much air and water they include), permeability (how fast water moves through them), and consistency. Even bedrock may be accorded different degrees of hardness, and the location of any fracture lines will be noted.

Underpinning

Skyscrapers are generally constructed in dense urban areas—and often directly up against adjoining buildings. Because digging a new or deeper foundation for a skyscraper will likely disturb the soil underneath adjacent buildings, these neighbors—big or small—will often require underpinning before construction of the foundation can start.

Underpinning counteracts or prevents subsidence of the soil and resulting settlement or collapse of the existing building. It also protects against "heave"— the upward pressure of the soil on a building.

Typically underpinning involves digging below an existing foundation and pouring additional concrete to extend either its width (to better distribute the existing load) or

its depth (so that the neighboring building rests on more solid material). A variety of techniques may be used, including the placement of temporary supports to "jack up" the building as excavation proceeds. Others involve the construction of additional foundation supports and then a choreographed transfer of the existing building from its original foundation to the new one.

Test borings, surveys, or regulations demonstrate the need for underpinning of buildings adjacent to a building site. The location of utilities must be clearly identified prior to excavation being undertaken.

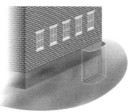

An excavation pit is dug to allow access to the neighboring building. It must be properly shored and strutted to prevent cave-ins.

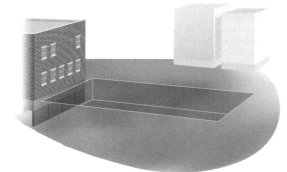

Full excavation takes place only after the neighboring structure is fully underpinned.

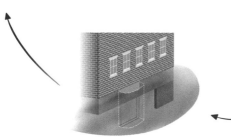

The process is repeated in a sequenced manner, determined by a structural engineer, to ensure the building's structure is not damaged.

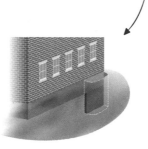

The building's existing foundation is strengthened through expanded concrete footings or by tying it into additional driven piles.

Whistles

Bedrock conditions often require more than just large excavators and backhoes. Moving the rock requires dislodging it first, generally through some form of blasting. Before any underground rock blasts occur, a series of whistles lets site workers and passersby—as well as workers and residents in nearby buildings—know that a blast is imminent.

In Manhattan, for example, one whistle means three minutes until blasting, two whistles means one minute until blasting, and a series of three whistles signifies an imminent blast. Following a blast, a long whistle will be given as the "all clear" signal.

1

A hole is drilled with a small rock drill or jackhammer anywhere up to 10 feet deep into the rock. This hole will hold the explosives.

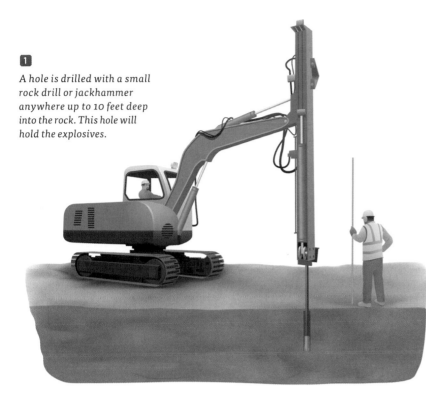

2

A blasting cap is inserted into the dynamite. The highly explosive cap will be used to initiate the explosion.

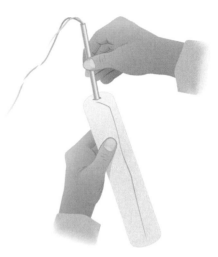

3

The explosive is lowered into the hole. This and any other charges are tamped down and then covered with dirt, sand, or clay.

4

Large steel-mesh mats are placed over the blast hole and surrounding area to contain fragments and concentrate the explosion. Each mat can weigh up to a ton.

5

After a series of safety checks are made and warning whistles are sounded, the "shooter" connects the blasting wires to the blasting box and detonates the charges.

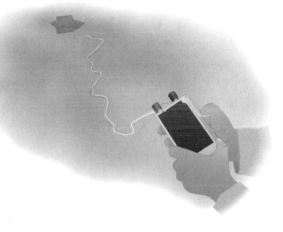

Blasting

Digging deep in the middle of a city often demands more than a powerful excavator. It may require the blasting of earth well below the surface of a job site, particularly if rock is found at or above the levels that will constitute the foundation. Various forms of dynamite will generally be used to remove it.

The placement of dynamite underground is done by drilling a careful pattern of blast holes and then wiring together the explosive charges placed inside them. A "blasting cap," made of pure nitroglycerine, is located at the end of a cartridge and placed at the bottom of the hole. Dirt or sand is packed in the hole on top of the cartridge and heavy, metal-mesh blasting mats are placed at the surface to protect site workers from any material sent flying by the explosion.

Once the dynamite is set and covered, an electric charge—initiated by a plunger on a remote blasting box—is carried through wires to the individual caps. The resulting explosions are timed to go off in a specific sequence—often seconds, or fractions of a second, apart.

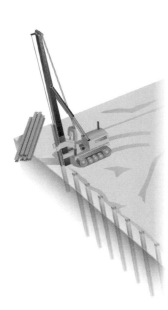

Digging a hole: the basic steps

1

PILES FOR SHEETING

A pile driver sinks a continuous perimeter of steel sheet piling into the ground.

2

EXCAVATION

Excavation proceeds within the area that is surrounded by sheet piling.

3

BRACING

Bracing, in the form of steel members placed inside the pit or tieback cables extending outward from it, is installed to further protect the integrity of the excavation.

4

DEWATERING

Depending on the level of the local water table, "dewatering" may be necessary during foundation construction. A variety of well and suction techniques help lower the amount of water flowing into the site.

Sheeting and Bracing

You don't need to be an engineer to know that the first step in constructing a foundation is digging a hole. But in the world of tall buildings, the earliest stages of foundation work must happen before excavation begins. Sheeting and bracing must be placed around the perimeter of the site to ensure that the soil around the soon-to-be-dug hole will not collapse into it during construction (like a hole in the sand at the beach) and that the foundation's perimeter wall, generally made of reinforced concrete, can be built precisely as designed.

A variety of technologies may be used to retain the earth around a hole. Often

sheet piles—thick sheets of steel woven together to form a continuous, interlocking barrier—will be driven one at a time into the earth prior to the start of excavation. Alternatively, soldier beams—wide-flange steel sections resembling the letter "H"—may be placed into the earth six to 10 feet apart, so that steel sheeting or wood may be inserted between them during excavation.

Once excavation is under way, steel sheeting alone is not enough to resist the pressure of the soil pushing into the newly dug hole. Various bracing methods are typically used to stabilize the sheeting.

Long structural elements may be placed across an excavation so that loads from the soil on opposite sides of a job site resist each other. Alternatively, bracing may be placed on a diagonal between the sheeting wall and the ground inside the excavation.

"Tiebacks," also referred to as "cables" or "stays," are often used to provide additional support. Tiebacks are steel tendons that are drilled into the wall face and anchored well beyond any area of potential soil instability. They exert pressure in the opposite direction of the soil outside the perimeter wall, further supporting the sheeting.

Machines in the pit

EXCAVATOR
The workhorse of the pit

DRAGLINE EXCAVATOR/BUCKET CRANE
Used to remove soil and rock

PILE DRIVER
Used to drive piles for the foundation

DUMP TRUCK
Used for transporting loose dirt, gravel, and rock

LOADER
Used to load loose construction debris into trucks

AIR COMPRESSOR
Provides high-pressure air to many handheld tools at the site

JACKHAMMER
Used for smaller, precision excavating

BACKHOE LOADER
Performs the function of an excavator and a loader

BULLDOZER
Used to push large quantities of debris

Diaphragm Walls

From time to time more substantial permanent walls may be constructed to assist the excavation process. Known as "diaphragm" or "slurry" walls, they assist with earth retention and prevention of water inflow during excavation, and can serve as permanent (though generally not load bearing) basement walls.

The first step in constructing a diaphragm wall is digging a trench around the site. To resist the earth pressure as the trench is dug, it is filled with a slurrylike mixture (generally made of water and an expansive form of clay) that expands and prevents water or earth from entering the trench. A steel frame composed of reinforcing bars is then inserted into the slurry.

Following insertion of the frame, concrete is pumped into the trench from its bottom. The concrete pushes the slurry upward, allowing it to be easily pumped out from the top of the trench. The concrete then hardens in place around the steel frame to form a permanent wall. Tiebacks may be inserted through it to further offset the weight of the soil pushing inward on the new wall.

1

Trenches are dug down to bedrock level. While digging is done, a slurry made of water and an expansive clay called bentonite is piped into the hole.

2

The bentonite slurry material expands along the sides of the trench, blocking the groundwater from seeping into the excavation pit.

3

A steel frame is lowered into the newly dug hole.

The World Trade Center's "bathtub"

The World Trade Center "bathtub" is perhaps the best-known example of a diaphragm wall. Because of the site's proximity to the river, the earth was saturated and did not lend itself to simpler excavation and dewatering techniques. A huge slurry trench, four blocks wide by two blocks long, was designed to serve as "the bathtub" to keep the river out of the site. Tiebacks were inserted in the wall prior to site excavation to ensure that the difference in pressure during excavation would not compromise the wall's integrity. The bathtub remained intact following the destruction of the buildings in 2001, and is now the site of the construction of the 9/11 monument and new office towers.

City of fill

Depending on the size of the planned skyscraper, enormous amounts of material may be excavated from one job site. In the case of the World Trade Center in the 1960s, roughly 1.2 million cubic yards (917,000 cubic meters) of earth and debris was excavated from the 16-acre (6.5 hectare) site to accommodate the complex's foundation. Instead of trucking or barging this material outside of New York City, it was used as fill to help create 23.5 acres (9.5 hectares) of land alongside the river's edge just opposite the Trade Center site (most of the other fill came from harbor dredge spoils). Today that fill is known as Battery Park City—and is home to thousands of residents and millions of square feet of commercial office space in addition to supermarkets, schools, museums, yacht marinas, and a highly acclaimed waterfront park.

MANHATTAN

■ WORLD TRADE CENTER
■ LAND ADDED (BATTERY PARK CITY)

4 *Concrete is poured into the bottom of the trench. At the same time, the bentonite slurry is pumped out of the top of the trench.*

5 *Tiebacks are driven through predetermined points in the wall to prevent the diaphragm wall from collapsing inward during construction.*

The fabulous Baker boys

There are two of them. They're both named Baker and both (appropriately) from Chicago, one of the world's greatest skyscraper cities. And while they're a generation apart and not in any way related, it is fair to say that—with the possible exception of Cesar and Rafael Pelli—they have together had more impact on the world of skyscrapers than any other name-connected pair in the business.

Clyde Baker, a geotechnical engineer by profession, knows as much about constructing foundations for tall buildings as almost anyone on the planet. And well he should; at age 80, his years of experience are unmatched. Seven of the 20 tallest buildings in the world were designed to stand on his foundations, and he has been involved as a consultant in countless others. Clyde calls himself "precomputer," but he is hardly old school: the innovative foundations that today support the Petronas Towers as well as the new Burj Khalifa in Dubai were designed with his input.

The "other Baker" is Bill, generally recognized as the foremost structural engineer working in skyscraper design today. As a partner at Skidmore, Owings & Merrill (SOM) in Chicago, Bill can trace his lineage back to Fazlur Khan—who held the same job at SOM when he devised his "framed tube" concept that would revolutionize skyscraper design. Baker's influence may ultimately be just as far-reaching as Khan's: his concept of the "buttressed core" allowed the Burj Khalifa to reach a full 1,000 feet (305 meters) higher than Taipei 101, the tallest building in the world between 2004 and 2009.

Caissons and Piles

There are far more words to describe technologies to transfer loads from a building to soil or rock than there are technologies themselves. The words "piers," "caissons," and "piles" are most commonly used, though their respective meanings vary from country to country. To confuse matters further, caissons are often referred to as "drilled piles."

To keep it simple, most foundations can be thought of as comprising either driven or bored piles. Driven piles are prefabricated structural elements driven into the ground. They compress the soil around them in the process, creating friction and thereby increasing their own load-bearing capacity.

Bored piles, often referred to as caissons, are often thread through soft material to reach rock or other bearing material. Typically they are fabricated on-site and sized to fill a predrilled hole in the ground. Temporary casing may be used to protect the hole while it is being drilled. Subsequently, a reinforcing cage will be placed in the completed hole before concrete is poured around it.

Piles come in many sizes and types and may be made of steel or a mixture of concrete and steel. Their configuration, including their length, is determined by an analysis of the soil in which they will be placed as well as by the physical characteristics of the construction site.

Piles are also differentiated by the way in which they transmit their loads. "End-bearing" piles carry their loads directly to soil or suitable bearing material deep under the building's foundations; consequently, they are only as strong as the soil or rock that lies at their base. "Friction piles," also referred to as "cohesion piles," transfer loads along their entire length through a form of skin friction. They are often driven in groups, to reduce the compression of the soil.

In addition to piles and caissons, many skyscraper foundations feature "mats" or rafts"—flat expanses of reinforced concrete used as building supports on shorter buildings or where rock is shallow. On supertall skyscrapers, these mats are often used in combination with piles or caissons—as a way of knitting them together and providing further stability to the building's foundation.

Driven piles

Driven piles are typically used when the underlying ground condition is granular or rock. Prefabricated piles are often made of prestressed, reinforced concrete or steel. The displacement that occurs as the pile is being driven compresses the soil around it, thus increasing both the friction against it and its load-bearing capacity. The length, shape, and material of the pile depend on the type of soil, the need to minimize vibration during construction, and the nature of access to the building site.

PILE FORCE
Every one to two seconds, a pile-driver can deliver up to 250,000 ft-lbs per blow. This is equivalent to dropping 30 to 45 Volkswagen Beetles from a height of seven stories.

PILE SHAPES
Reinforced concrete–driven piles will typically be shaped like a nail, but steel piles may take the shape of an X or H cross section or a hollow pipe.

Pile testing

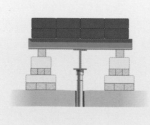

Through pile testing, it is possible to check that the piles as designed will be adequate to support the building's weight before completion. In a Kentledge pile test (shown), reaction weights are stacked on top of a structural grillage that distributes weight over the piles; the applied force versus the displacement of the tested pile is measured.

Bored piles

Bored piles, also referred to as nondisplacement piles, drilled piers, or caissons, are generally used in softer or claylike soil conditions. A shaft is drilled by a drill rig, typically relying on pneumatic, hydraulic or diesel power and employing varying types of drill bits. The drilling of the shaft and the installation of the pile, usually consisting of a metal reinforcing cage and cement, are generally vibration-free.

BORING A PILE

1 *An auger is inserted into the ground with a drill rig. A steel casing or bentonite mud may be necessary to support the new hole.*

2 *When the required boring depth is reached, a reinforcement cage is installed and the pipe is filled with concrete.*

3 *Finally, the auger and supports are removed.*

DRILL BITS

Hollow-stem augers are used for loose soil, clay, and surface materials.

Farther down the hole, hammer bits are used for soft rock, like calcites, dolomites, limestones, and hard shale; the drill string is rotated along with pneumatic pressure, and nibs on the bit shear the rock.

Rotary drill bits consist of teeth on wheels that turn as the drill string is rotated; these teeth apply a crushing pressure to the rock, breaking it up into small pieces.

The deepest foundation

CROSS SECTION

VIEW FROM ABOVE

The foundations of the Petronas Towers are the world's deepest—extending roughly 500 feet (150 meters) below grade. Because the limestone bedrock below the site slopes sharply, foundation engineers initially designed a hybrid foundation system: one side of the complex would rest on bedrock and the other on concrete columns. But analysis suggested that such a mixed foundation system might settle unevenly—awkward for two buildings connected at the top. So the building was relocated 200 feet (60 meters) east, to an area that could evenly support a thick concrete mat connecting bored piles.

The piles chosen for the foundation were unusual: they were rectangular, not circular, and featured indentation along their sides. Known as "barrette piles," their novel shape increased friction with the surrounding soil. During their placement, however, the limestone at the base of the piles was found to be unacceptably porous. As a result, reinforcing grout was injected up to 100 feet (30 meters) below the top of the bedrock layer to hold it together permanently.

Earthquakes

Foundations for skyscrapers are tailor-made and take account of the properties of the soil at a particular site as well as the specific design and weight of the building. Nowhere is this more true than in an earthquake zone, where a tall building is expected to withstand several major seismic events over the course of its life.

The vulnerability of buildings to earthquakes largely relates to their "natural frequency"—the degree and speed at which a building will move back and forth in response to a seismic wave. Buildings vary widely in their natural frequencies due to height, structure, and stiffness. In general, those buildings that happen to move back and forth in sync with a seismic wave will greatly amplify its force upon them—a phenomenon known as "resonance"—and will often be damaged or destroyed in the process.

A variety of foundation and structural techniques have been devised to protect buildings from seismic events. In smaller structures, a variety of specialized interconnections, generally referred to as "base isolation techniques," are used to permit a building to slide around on its foundation—thus minimizing the transfer of seismic energy to the structure above. But because tall buildings can suffer "uplift" problems (water from below or lateral forces can push a building upward), a stronger connection between tower and foundation is required.

Foundation technologies used to stabilize tall buildings in areas of frequent seismic activity include reinforced concrete mats (also called "rafts") that sit under the structure and connect the piles or caissons drilled deep down into the earth. These mats can be as large as six feet (1.8 meters) thick and the piles and caissons they connect will generally be long—extending down to stable, load-bearing soil.

Above the ground, other technologies are often employed to minimize the building's movement. Fluid, viscous dampers—a form of building shock absorber—are increasingly used within the steel structure of tall buildings to dissipate seismic energy. On the Torre Mayor in Mexico City, for example, these dampers can be seen running alongside the external bracing (another feature commonly used to minimize a building's lateral movement) on the building's perimeter.

The pyramid's secrets

No area of the United States is as earthquake-prone as California, and San Francisco—located atop the San Andreas and Hayward faults—is particularly prone to them. As a result, the city's skyscrapers are built to resist earthquakes.

The Transamerica Pyramid, perhaps the world's most recognizable modern pyramid, is a good example. Located atop land that at one time was part of San Francisco Bay, it required a particularly strong and deep foundation. To support its 48 floors, the building's concrete mat foundation—comprised of some 16,000 cubic yards (12,200 cubic meters) of concrete and 300 miles (483 kilometers) of steel reinforcing rods—was sunk 52 feet (16 meters) into the earth and designed to move with the tremors.

Aboveground, its quartz cladding was designed with clearance between its panels to allow for movement of the skin during a seismic event. To provide additional support for horizontal and vertical loads, a unique horizontal truss system was installed above the building's first floor.

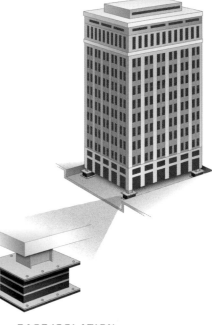

BASE ISOLATION
While typically not used in tall skyscrapers, a base-isolation system is one of the most commonly used methods of seismic protection. The principle of base isolation involves decoupling the building from ground motion during an earthquake by mounting rubber bearings between the building and its foundation. Pasadena and San Francisco city halls and several of Salt Lake City's municipal buildings have been retrofitted with base-isolation systems.

North American earthquakes

As shown on the map, the western half of North America is prone to larger earthquakes, due to several major fault lines that snake through the Pacific Northwest, California, and Mexico. A number of techniques are used to mitigate their impact on the built environment.

SEVERITY OF EARTHQUAKES

▪ MOST SEVERE
▪
▪
▪
▪
▪
▪ LEAST SEVERE

Resonance

Like striking a tuning fork, an earthquake's unique energy will cause buildings of certain heights to experience extreme motion, or "resonance." This typically occurs when the frequency of the building's vibration (or "natural frequency") matches the vibration of the earthquake. Sixty percent of the buildings damaged in Mexico City's devastating 1985 earthquake were between six and 15 stories; those taller and shorter were less likely to experience damage.

▪ EARTHQUAKE VIBRATION
▪ BUILDING VIBRATION

BUILDING HEIGHT (IN STORIES)

30
20
10
0

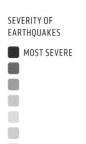

VISCOUS DAMPERS

Viscous dampers, a form of structural shock absorber, can reduce the movement of structures in an earthquake by as much as half. They are typically placed in a diagonal or chevron configuration between structural columns of a building. Buildings relying on these dampers include Los Angeles' city hall, the San Francisco Civic Center, the Tokyo-Rinkai Hospital, and the Torre Mayor in Mexico City.

TORRE MAYOR

STRUCTURE

The place to begin understanding skyscraper structure is in Egypt—just outside of Cairo. The pyramids of Giza have stood there for 4,500 years—the longevity of their enormous bases and tapered tops a credit to a man named Imhotep, often acclaimed as the world's first structural engineer. He calculated correctly that a thick, strong base was a critical ingredient in building tall and, relying on that premise, built pyramids that reached higher than any structure preceding them.

Imhotep's simple building formula remained unchanged for thousands of years. Until the late nineteenth century, multifloor structures around the world were supported on masonry walls that increased in thickness as they neared the ground. The higher they reached into the sky, the thicker their bases needed to be. This placed an effective cap on the height of buildings, as at some point floor area at the base became so monopolized by supporting walls that it was useless for any other purpose.

The idea of a skeletal frame—an internal frame of metal used to support the floors and walls of a tower—was a radical one when espoused by William Le Baron Jenney in 1884, upsetting thousands of years of building history. But its merit—particularly given the declining cost of steel—was quickly recognized, setting off the race for height that would mark the dawn of the age of the skyscraper.

The primary components of a skeletal-frame system are columns and beams. Together they transmit horizontal and vertical loads through the building while supporting secondary elements such as floor slabs and cladding. Columns run vertically and transmit gravity loads to foundations; beams are aligned horizontally and transfer gravity loads to the columns. Secondary steel beams, or girders, support the floor structure.

This new rigid-frame system served as the backbone for nearly all skyscrapers built during the first half of the twentieth century. The earliest skyscraper designers saw merit in highlighting the rectilinear nature of this system on the building's exterior, but by the second decade of the century frames were increasingly hidden by ornamental masonry façades that added weight, but no support, to the structure. Structures like New York's Chrysler and Woolworth buildings and Chicago's Tribune Tower relied on a skeletal-frame system that is largely invisible from the outside.

The rise of structural expressionism in the mid-twentieth century marked a "coming out" for the steel skeletons that had, to that point, been hidden beneath masonry walls. The use of glass façades allowed the boxy shape of the skeletal frame to be seen and, in the case of architectural masterpieces such as the Seagram Building and Lever House, celebrated.

The economics of steel

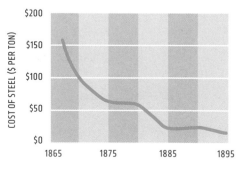

The skeletal-frame system

Load-bearing wall buildings, common until the late nineteenth century, had walls up to six feet (1.8 meters) thick at street level and very small window openings.

The new, steel-framed buildings featured shallow window depths and more window area. They also provided more usable space on the lower floors of the building.

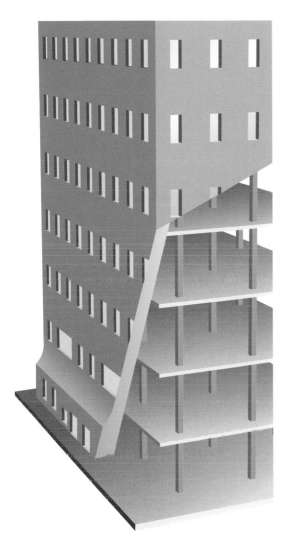

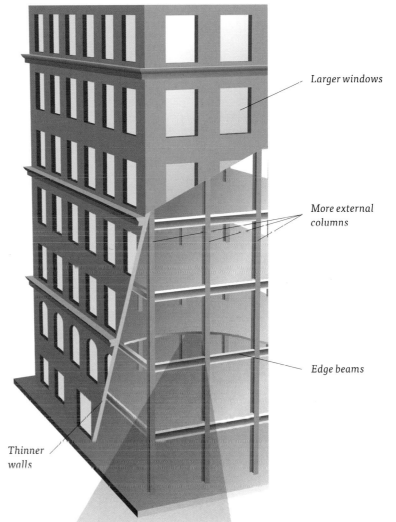

Larger windows

More external columns

Edge beams

Thinner walls

COLUMNS
Vertical elements that transmit gravity loads from the floors to the foundations

BEAMS AND GIRDERS
Horizontal elements that support the floor and transmit their vertical loads to the columns

FLOOR SLAB
Concrete-based system that resists the vertical loads directly acting on it and transmits these loads to the supporting floor beams and girders

FOUNDATION
Structure designed to transmit vertical loads from the columns to the earth

How tall should it be?

Over time, the structural systems supporting skyscrapers have changed dramatically. Each system, beginning with the earliest rigid steel frame and continuing through today's buttressed core, has a "maximum efficient height," above which the form becomes less than optimal from a structural perspective. However, because commercial viability is as important to developers as structural integrity, not all skyscrapers have been built to their efficient height.

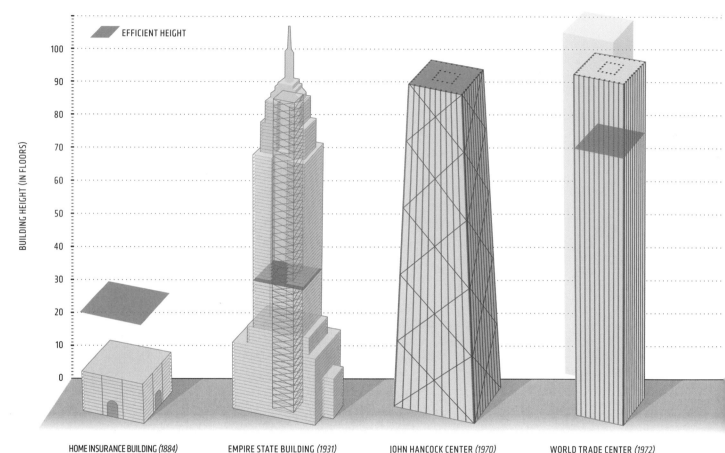

EFFICIENT HEIGHT

BUILDING HEIGHT (IN FLOORS)

HOME INSURANCE BUILDING *(1884)*

EMPIRE STATE BUILDING *(1931)*

JOHN HANCOCK CENTER *(1970)*

WORLD TRADE CENTER *(1972)*

RIGID-STEEL FRAME
The earliest skyscraper structural system had an efficient height of somewhere between 20 and 30 stories.

BRACED RIGID FRAME
Bracing was added to resist lateral loads as buildings got taller, yielding an efficient height of 40 stories.

BRACED TUBE
The lateral load-resisting system was moved to the building's perimeter—pushing the efficient height to over 100 floors.

FRAMED TUBE
The framed-tube design of the World Trade Center towers, which relied on closely spaced columns, had an efficient height of 80 stories.

Structural Evolution

The rigid-frame system was not exclusively a steel phenomenon. Concrete was used to frame tall buildings as early as 1903, but they were bulky affairs and rarely reached taller than 15 stories. Steel frames could reach three or four times higher—until wind loads became so great that they would jeopardize the building's structural integrity.

Building higher than 50 or 60 stories became possible only with the development of an entirely new framing system in the mid-1960s. Tubed systems, the idea of

a structural engineer at Skidmore, Owings & Merrill named Fazlur Khan, relied on a series of perimeter tubes designed to both carry the gravity loads of the building and resist the lateral loads upon it. Tried first on a modest apartment complex in Chicago, the idea would revolutionize skyscraper design around the world.

Perimeter-tube systems, expressed in either concrete or steel, relied on closely spaced exterior columns that, connected to one another by deep beams, formed a

spatial skeleton for the building. Tubes could be arranged in any form whatsoever, freeing buildings from the rectilinear shape that had characterized rigid-frame systems. The strength of the perimeter structure meant that far fewer internal columns were needed, opening up the interior of tower floors dramatically.

The increased stability associated with the new tube technology allowed buildings to resist greater loads and thus rise to unprecedented heights. The four giant

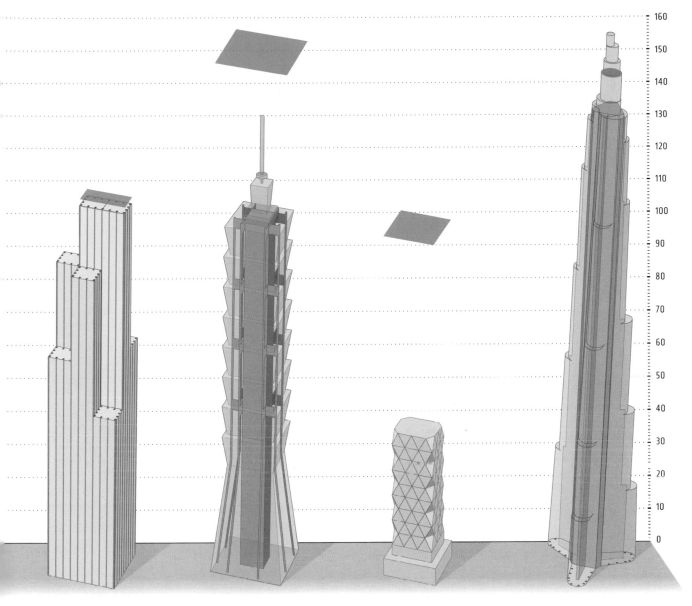

BUILDING HEIGHT (IN FLOORS)

160
150
140
130
120
110
100
90
80
70
60
50
40
30
20
10
0

WILLIS (SEARS) TOWER *(1974)*

TAIPEI 101 *(2004)*

HEARST TOWER *(2006)*

BURJ KHALIFA *(2010)*

BUNDLED TUBE

The bundled-tube system relied on a cluster of tubes connected together to act as a single unit; it allowed better interior planning and a slightly higher efficient height.

CORE AND OUTRIGGER

The conventional load-resisting system is extended from the core to the building's perimeter columns through the use of outriggers, raising the efficient height to 150 stories.

STEEL DIAGRID

The steel-diagrid system resists lateral shear via its diagonal members. Though harder to construct due to complicated joints, it has an efficient height of roughly 100 stories.

BUTTRESSED CORE

The buttressed-core system, devised for the Burj Khalifa, features a sexagonal core reinforced by three buttresses that form a Y shape and allow the building to support itself.

towers of the 1970s were all tube-based systems: the World Trade Center towers were framed-tube structures, Sears Tower was designed as a bundled-tube system, and Chicago's John Hancock building was designed with a braced-tube exterior. Though their bracing systems varied, each featured lateral load-resisting tubes at the building's perimeter.

While the views from the upper floors of these new supertall buildings were unmatched, not all portions of these floors

enjoyed them. The perimeter tubes blocked views to varying extents: one third of the Sears Tower façade was devoted to the structure, while half of the World Trade Center façades were made up of steel rather than glass.

A new structural system was devised in the 1990s that improved upon the tube idea. In place of myriad vertical columns the building perimeter would feature a smaller number of supercolumns attached to a high-strength concrete core

by massive outrigger arms that stretched horizontally across the building at various levels. The supercolumns, like the tubes before them, were attached to one another by belt walls that encircled the perimeter at regular intervals.

Over a period of 100 years, then, the structure of the skyscraper really became its architecture. Designed to work with both gravity and the wind, the modern skyscraper found itself climbing to heights never before imagined.

Steel

The production of steel involves the melting of iron ore and the addition of various other elements, often called alloys. The mix of these alloys determines not only the hardness of steel but other properties as well. For example, the addition of chromium leaves a hard oxide on the surface of the steel, giving us what we know as "stainless steel." In general, the different grades of steel are characterized by their "yield strength"—the stress level at which a steel member begins to bend permanently out of shape.

Ways of increasing the strength of steel have been devised over the last century, so that today a thinner steel member can provide a level of support that required a much thicker one at the dawn of the skyscraper age. Plants around the world distinguish themselves by the size and thickness of the steel they can produce, with those producing the largest steel plate generally located in Europe and Asia.

1

The first step in steel formation is a blast furnace. Fuel and iron ore are continuously supplied through the top of the furnace, while air is blown into the bottom of the chamber. The furnace produces molten metal and slag.

2

The molten metal and slag is poured into a ladle or torpedo ladle. The high-strength, high-temperature ladle is used to transport and pour the molten metal for the casting process.

3

Continuous casting transforms molten metal into a solid on a continuous basis. In this process, molten steel flows from a ladle through a "tundish" into the mold. The tundish can also serve as a refining vessel to float out impurities into the slag layer. After the mold the cooling steel is passed through withdrawing rollers.

4

The steel reheat furnace is used to raise the temperature of the steel to 2,200 to 2,400 degrees Fahrenheit (1,200—1,300 degrees centigrade) in order to "hot roll" the steel in a desired shape and thickness. The yield strength of steel decreases as the temperature rises, and large deformations can thus be obtained with modest roll forces.

5

The hot steel is passed through a rolling mill to shape the steel, using rollers configured in different axes to produce shapes ranging from flat plates to I-beams and C-channels.

Concrete

Concrete is a construction material made up of cement, aggregates (normally a coarse material like gravel, limestone, or granite, plus a finer additive like sand), water, and various chemical admixtures. Use of concrete dates back to ancient times, when lime was a common element in mortar. Its composition has varied greatly over time, and added ingredients have included materials as wide-ranging as blood (for frost resistance), hair (to reduce cracking), and volcanic ash (to accelerate hardening).

The use of concrete in skyscrapers was limited until the introduction of reinforced concrete—i.e., concrete strengthened by the insertion of steel. Though it performs well under compression, concrete performs poorly under tension. Steel, in contrast, performs equally well under either condition. Combining the two initially took the form of inserting straight steel bars into concrete (with metal stirrups to clip them on); soon after, twisting steel bars were introduced to increase the adhesion of the steel to the concrete itself.

While reinforced concrete was used heavily in the construction of multistory residential buildings (generally 10 to 20 stories) through the first half of the twentieth century, it was only with the development of new forms of concrete in the second half of the 1900s that concrete came into its own as a preferred material in skyscraper construction. Stronger steel was introduced to reduce the weight, and hence increase the efficiency, of tall towers. New forms of high-strength concrete allowed tower columns to be designed thinner than previously. And the introduction of high-performance concrete, which cured, or reached its strength, faster than previous forms of concrete, vastly sped up construction on building sites using it.

Today the advantages of using these two materials in tandem has been further amplified by the practice of casting concrete around pre-tensioned steel tendons. The cured concrete, undergoing a chemical reaction, bonds to the steel tendons; when the tension is released it is transferred to the concrete as compression by static friction. Most pre-tensioned concrete elements are prefabricated in a factory and transported to the construction site by truck—thus limiting their size. Similar results can be achieved by "post-tensioning concrete," or by adding tension to the steel tendons after pouring the concrete.

Steel-reinforced concrete

While concrete is relatively cheap and can easily take the shape of any form, it shows strength only when compressed. If a concrete structure is pulled apart, it wants to crumble. Steel, on the other hand, has the same strength whether compression or tension is applied. Reinforcing concrete by embedding steel within it improves the behavior of concrete in tension by taking advantage of steel's strength in both directions.

The term "rebar" is commonly used as shorthand for "reinforcing bar"—the cylindrical steel bar that is used to reinforce concrete. Often the rebar will be ridged, in order to adhere better to the concrete. Regardless of its form, however, rebar can still be separated from the concrete in high-stress conditions. As a result, it may be bent and hooked to other rebar at its end—to lock it more securely into place.

Structural steel used in buildings typically has a yield strength of approximately 50,000 pounds per square inch (psi), but higher-strength variants may deliver strengths over 100,000 psi.

High-strength concrete mixtures (nonreinforced) can deliver compressive strengths around 10,000 psi; however, under tension, these same concrete mixtures may only exhibit 10 to 15 percent of their strength in compression.

Reinforcing concrete can significantly improve its performance. The addition of high-strength steel or carbon-fiber reinforcement, as well as the incorporation of quartz or silica aggregate, can help concrete mixtures reach compressive strengths in excess of 100,000 psi.

Mixed-Use Structures: Concrete and Steel

Until the late twentieth century, the structure that supported tall buildings was made from either steel or concrete. Pre-1900 skyscrapers generally relied on steel frames, while the introduction of reinforced concrete at the turn of the century offered builders another option. In general, developers of commercial buildings chose steel and those developing residential buildings chose concrete.

Concrete was the material of choice for residential construction for a variety of reasons. It is far better than steel at sound absorption, and reducing the transfer of

noise between individual units of a hotel or apartment building was an important consideration to developers of these buildings. It is also more fire-resistant, an important characteristic in an age when fires were a frequent occurrence and fire prevention technology was more primitive than it is today.

But concrete has limited floor spans—not ideal for office buildings, in which larger, more flexible floor plates are desired (and proximity to a window is less important than it is in homes). Both commercial developers and their tenants have always

put a premium on open, column-free space—which can more easily be adapted to the needs of a variety of office users over time. Steel, which can support much longer individual spans than concrete, was therefore the material of choice for office construction throughout most of the twentieth century. The demands of modern technology—e.g., underfloor air-conditioning, telecom, and power—highlight the continuing value of flexibility and openness in the design of tower floors.

Advances in concrete technology have, however, blurred the relative advantages

The rise of the mixed-use building

Historically speaking, most of the tallest skyscrapers in the world were office buildings. This remained true as recently as the year 2000. Over the last decade, however, there has been a significant shift toward residential uses. Today less than half of the 100 tallest buildings in the world are used solely for office purposes.

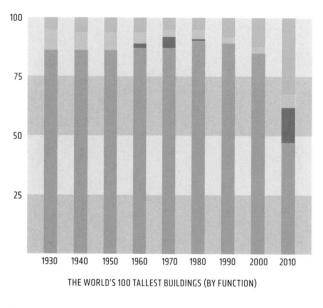

THE WORLD'S 100 TALLEST BUILDINGS (BY FUNCTION)

- ◼ MIXED USE / OTHER
- ◼ HOTEL
- ◼ RESIDENTIAL
- ◼ OFFICE

Anatomy of a hybrid

A mixed-use tower, such as the Bloomberg Tower in New York (also known as 731 Lexington Avenue), generally employs steel framing for the lower, office portion of the building, to ensure maximum spans from column to column. A transfer floor is then built to support the residential tower above, which features a shorter distance between columns and is framed in concrete to ensure better sound-proofing. The distance between columns in the office portion of the Bloomberg Tower is 30 to 35 feet (9 to 10.5 meters); the distance between columns in the residences is 15 to 20 feet (4.5 to 6 meters).

Bloomberg Tower under construction

of the two materials in skyscraper construction. The development of high-strength and high-performance concrete has meant that concrete buildings can be designed with skinnier columns and can go up much faster than ever before—both helping to offset the cost of construction. In addition, new technologies allow concrete to be pumped higher and higher. In 1960, the tallest concrete building was 20 stories; 50 years later, concrete was the material of choice for some of the world's tallest buildings—including the Petronas Towers and the Burj Khalifa.

The ability to build high with concrete has facilitated the rise of the mixed-use building—one that serves both office and residential purposes. Indeed, many of the world's tallest structures—and even smaller ones, like the Time Warner Center in New York—now seamlessly blend the two uses along with many more, including retail, recreation, and entertainment. Office and retail spaces are usually found in the lower portion of a building, with hotel or residential units—both of which lend themselves to smaller floor plates and better views—located toward the top.

The construction of mixed-use buildings is an art in itself. Typically steel will be used to provide the long spans necessary for the office and retail portions of a building, with concrete used to construct the residential or hotel portions of the structure. Because the design and relative sizes of the floors vary dramatically from one to the other, a transfer floor will generally be needed to shift the loads from the columns in the top (typically concrete) section of the building to a new grid of columns located in slightly different places in the lower (typically steel) portion.

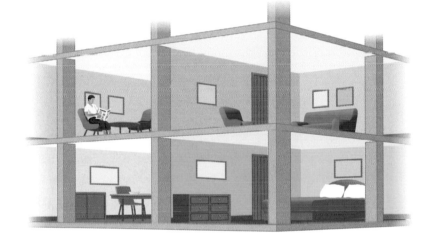

RESIDENTIAL (CONCRETE) FLOORS
Concrete-framed structures in residential skyscrapers often have much closer column spacing than steel-framed buildings. Shorter spans between columns do not present as much of a problem with interior layouts, since residential floors are divided into much smaller rooms than typical office floors.

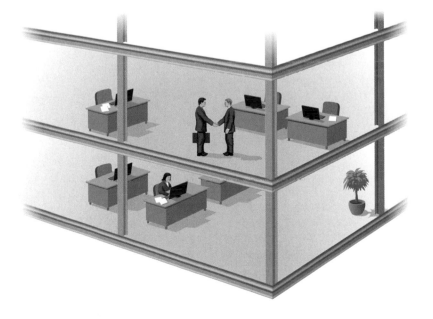

OFFICE (STEEL) FLOORS
Steel-framed office structures allow much longer unsupported spans between columns—typically up to 45 feet (14 meters). This, in turn, affords the architect and tenants more flexibility in laying out office floors by offering larger, open spaces or ample room for division into offices by non-load-bearing walls.

Gravity Loads

The structural system chosen to support a given building is primarily a function of the loads to be placed upon it, and will almost always be designed in conjunction with its foundation. A variety of loads are taken into account by structural engineers during the design process, including the dead load—the weight of the building alone—and the live load—the weight of the people and the furnishings that will populate the structure.

Dead and live loads are important in calculating what's known as the "gravity load"—the weight pushing down from the building into the earth below it. This is sometimes referred to as the "vertical load," although this phrase technically includes loads pushing upward (sometimes called "heave" or "uplift") that arise as a result of either wind moving against a building's roof or groundwater or seismic energy pushing the entire structure upward from the soil.

Gravity loads are transmitted through a building's columns, which may be located on its perimeter or internally, including through its core. Typically each column will land on a footing, which acts to spread the gravity load across an area wider than the base of the column. These may often be attached to a raft structure, which often acts as the top of the piles, caissons, and other foundation elements.

Height and gravity

As the height of a building (and the amount of its "dead load") doubles, the total gravity load roughly doubles as well.

Floor slabs take dead and live loads on the floor and transmit them to floor beams.

Floor beams transmit these dead and live loads to vertically oriented columns.

Columns transmit these gravity loads to the foundation.

Transfer beams and floors

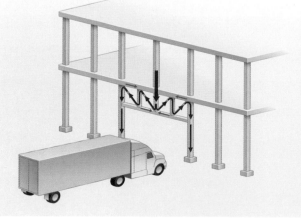

Gravitational forces have changed little since Sir Isaac Newton's day: they still pull items straight down to the earth. Building loads from the upper floors of a tower will typically travel straight down a series of columns directly to the building's foundation—unless they are interrupted.

For any number of reasons, regular column spacing may not be desirable on a particular floor of a building. There may be a need

for more open space—e.g., in a lobby or auditorium area—or there may simply be a desire for a less cluttered appearance on a particular floor.

But the all-important columns can't be interrupted or removed without making sure that the load they are carrying has somewhere to go. In cases like these, transfer beams—horizontal members spanning the roof of the space to be opened—will be used to

Lateral Loads

Few people outside the engineering or architectural professions realize that horizontal, or lateral, loads, rather than vertical ones, are what really drive skyscraper design. As building height grows, the lateral loads upon a structure grow exponentially, while the gravity loads simply increase arithmetically. The biggest challenge a designer of a supertall building faces, therefore, is how to brace the building to withstand lateral loads.

Lateral force upon a tall building can take several forms. It can arise from a seismic event, which sends shock waves out from its epicenter. Most commonly, however, it stems from the force of the wind as it moves past a particular building. Wind pressure is determined by many things, including the wind's speed and direction, the geometry of the building, and the shape and proximity of the structures upwind of the building.

Wind itself is rarely uniform; it is composed of numerous eddies of varying sizes and behavior as it moves past the earth's surface and past nearby structures and topography. As a result, its force can vary dramatically over fairly short distances and time intervals. In general, its speed generally increases significantly with height, though it also tends to behave more predictably (i.e., fewer gusts) at higher levels.

Assessing the wind is critical to building engineers. Too much movement could undermine the integrity of the beams and columns and lead to deformation of the structure, possibly causing damage to building components and discomfort to building occupants. The skin of the building, known as "cladding," is vulnerable to sudden gusts of wind, because it is not load bearing and is attached fairly simply to the structural elements beneath it. In addition, wind loads at the base of a tower can have a significant impact on people and objects at street level, creating an uncomfortable, and sometimes dangerous, situation for pedestrians.

Calculating wind loads is the law in many places. However, these regulations tend to assume simple rectilinear structures and do not take into account the dynamic nature of wind behavior, putting the onus on the design team to undertake more sophisticated modeling.

Height and wind

As a result of a doubling in height, the total lateral load acting upon a building will increase by a factor of four—or by the amount of the increase squared. As buildings get taller, wind loads increase much more significantly than gravity loads.

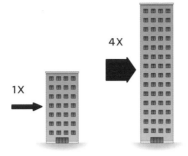

1X   4X

Wind loads act on the surface of the curtain wall.

The curtain wall is attached to the floor beams, which absorb the load.

The floor beams transmit the load to vertical columns or through cross-bracing.

An imbalance of lateral loads between the upwind and downwind faces of the building results in an overturning, or uplift force, on the windward columns and foundation.

shift the load from the columns above the floor to others on the edges of the open space. The size of the opening, or column-free space, will generally dictate the thickness and strength of the transfer beam required.

The concept of transferring load can also be applied on a broader scale to the entire weight of a building. When a structural engineer needs to shift the footprint of all the columns—either because the size and layout of the floors are changing or because spacing between the columns needs to change—he or she will design a "transfer floor," a trusslike structure that carries the gravity loads from the columns above to a new set of columns below the transfer floor. This is often the case with mixed-use buildings whose floor layouts change significantly between designated commercial and residential floors.

Wind Design

To minimize building costs and maximize revenues, skyscrapers generally take the shape of squares or rectangles. These are not particularly aerodynamic shapes, and their bluntness results in a wind phenomenon called "vortex shedding."

As wind meets a rectangular skyscraper it pushes on the flat face of the building before flowing around its sides, where eventually it separates from the face of the structure. The difference in pressures on the front and back faces of the building gives rise to vortices, or spinning eddies of wind, that flow downstream from the building. The vortices pull and push the building in a direction perpendicular to the wind at frequencies that can become dangerously self-sustaining.

To minimize vortex shedding, skyscraper designers do everything in their powers to confuse the wind. Orienting the building so that the longer face of the structure is parallel with prevailing winds can help. Chopping or rounding off corners of a building can likewise make it more aerodynamic. Roughing up the corners of the building through the careful placement of balconies or stepped corners can help by disturbing or delaying the formation of strong vortices.

Bigger architectural statements can also address the problem. Placing a twist in a building can "confuse" the wind by preventing resonant vortex shedding along a building's height. Likewise, rotating a building can minimize the wind load from the prevailing direction. For example, the curvilinear nature of the Swiss Re ("Gherkin") Tower in London generates pressure differences at its face that minimize wind loads. Punctuating the building's surface with an opening, such as the Shanghai Financial Center's trapezoidal void, can also serve to minimize the force of the wind upon a building.

To determine the best building configuration in a particular location, comprehensive wind-tunnel testing is done early on in design. The testing determines the strength and behavior of the wind both at the building site and in the areas around it, particularly at street level. It can also provide other useful information, such as the likely concentration of exhaust pollutants in the building's air intake system.

Wind-tunnel testing is an iterative process. Tall building designs are refined and tested repeatedly on the basis of test results—to the point where some skyscraper engineers now refer to their buildings as having been "designed in a wind tunnel." Far from a nice to have set of data, the results of wind-tunnel testing represent an integral part of modern skyscraper design.

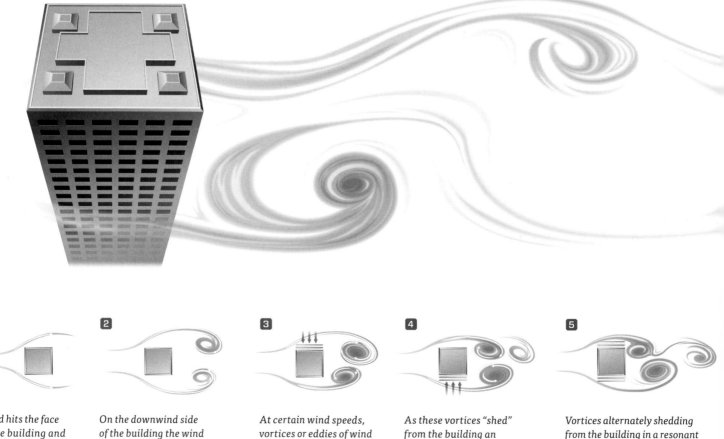

1 *Wind hits the face of the building and proceeds to flow around its sides.*

2 *On the downwind side of the building the wind flow separates from the building face.*

3 *At certain wind speeds, vortices or eddies of wind are generated.*

4 *As these vortices "shed" from the building an imbalance in forces occurs on the sides of the building.*

5 *Vortices alternately shedding from the building in a resonant condition can cause extreme side-to-side motions and forces on the building.*

Aerodynamics

Aerodynamics are a major consideration in the design of supertall skyscrapers today. Structural engineers will test many shapes and configurations to determine the optimum skyscraper design. Streamlined shapes, as well as twisting forms and rough or obstructed corners, can all serve to limit vortex-shedding behavior that can undermine the integrity of tall towers.

STREAMLINED SHAPES

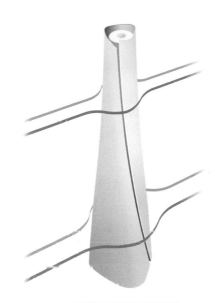

TWISTING FORMS

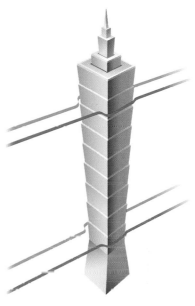

ROUGHNESS AT CORNERS

How a wind tunnel works

Most municipal building codes address permissible wind loads on new buildings. But because they are based on generic-shaped buildings, under generic conditions, they often fail to account for the unique aspects of individual tower designs—including the aerodynamic impacts of the building's shape, the effect and influence of adjacent buildings and local topography upon the wind, and the dynamic effects of wind directionality and localized airflows.

For this reason, building codes permit and often encourage wind-tunnel tests to be performed to evaluate a proposed building's performance under actual design conditions. The costs of such a study, while not insignificant, pale in comparison to the savings in cladding and structure that often result—sometimes in the millions of dollars. In addition, wind-tunnel testing confirms that the architect's vision can be safely realized, and thus offers protection against future litigation.

The first step in wind-tunnel testing involves construction of a replica of the building being tested and all buildings within a half-mile radius of it, typically on a 1:400 scale. The replica is placed on a rotating platform in a computer-controlled tunnel that features a fan capable of simulating a 60 mph (100 km/hr) wind. Thousands of sensors are placed on the model.

Wind flows that mimic natural wind in the area, including a realistic distribution of gust speeds, are then directed at the model. As the direction, angle, and strength of the wind changes, computers record both wind resistance on the building and wind conditions at the bases of it and other buildings in the area. Smoke is often used to make wind vortices easier to see.

Dampers

In many tall buildings mass dampers located near the top of the building act as a pendulum to shift weight around to counteract the forces of the wind against a building.

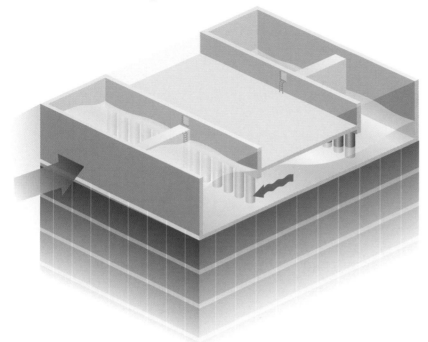

TUNED LIQUID COLUMN DAMPERS.

The Comcast Center in Philadelphia employs the world's largest tuned liquid damper. Holding 300,000 gallons (1.1 million liters) of water, its U-shaped tank allows the water to oscillate freely at the natural frequency of the structure.

TUNED MASS DAMPERS

While most dampers are hidden somewhere high in the building, the 728-ton (660-metric-ton) golden tuned mass damper of Taipei 101 swings in full view of patrons of the building's restaurants and observation deck on the eighty-fifth, eighty-sixth, and ninety-first floors.

Fighting Sway

Wind loads are a problem for skyscraper designers, but not necessarily for structural reasons. Movement back and forth at high levels rarely endangers a building's steel or concrete structure, but it does have the ability to induce motion sickness in humans living or working on those floors.

How far buildings are allowed to move back and forth is typically a function of how much movement people can tolerate—i.e., at what point a building's sway causes discomfort to its inhabitants. In the United States, it is common for tall buildings to be designed to sway no more than 1/500th of their height. Using this criterion, the new 1 World Trade Center—under construction in lower Manhattan—

would be permitted to sway just over three feet (one meter) at the top of the building.

A variety of techniques are used to limit the sway of a skyscraper. Bracing, often used to tie together the building's vertical and horizontal structural members, can take many shapes and sizes: K-bracing and V-bracing can commonly be seen on the outside of some skyscrapers. In areas of seismic activity, fluid viscous dampers often supplement bracing as a sort of shock absorber, dissipating seismic energy and reducing building displacement.

But really tall buildings demand more than just bracing and internal shock absorbers. Increasingly they rely on what are known as "mass dampers"—large

structures placed at or near the roof of a building that function like pendulums to shift weight in the opposite direction from the way that the wind is pushing the building.

The earliest mass damper consisted of a large hydraulically driven weight called a "tuned mass damper." The Citicorp Building in New York, at 59 stories, was one of the first buildings in the world to feature one. Subsequently, "sloshing" or "liquid" dampers were developed. Large tanks holding thousands of gallons of water, they allow the weight of the water to push in the direction opposite to that of the building's movement—naturally counteracting its motion.

Bracing

Lateral bracing provides stiffness to the building and resists movement from wind and other lateral loads. Braced-bay systems are the most visible type of lateral bracing. They comprise diagonal, cross, K, and eccentric bracing arrangements. Other systems include braced cores, shear walls constructed from reinforced in-situ concrete, and rigid, jointed, moment-resisting frames.

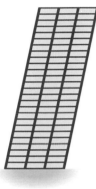

Moment frames feature extra steel welding of the beam-column connector along the exterior perimeter of the building to create a composite frame.

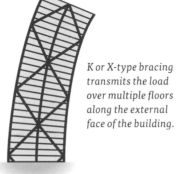

K or X-type bracing transmits the load over multiple floors along the external face of the building.

The belt truss system, featuring one or more belt trusses (sometimes referred to as a "hat" when atop the building), is a standard bracing system.

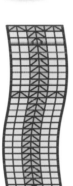

Belt trusses may be combined with a braced-core frame, which provides additional stability along the building's core.

Today dampers of one sort or another are more common than ever before as a part of tall-building technology. The new Comcast Building in Philadelphia features a "tuned liquid column damper" that features water moving in a column—rather than a tank—configuration. One Rincon Plaza, a new residential building in San Francisco, features a "tuned liquid mass damper" that comprises two 5,400 gallon (20,440 liters) tanks. And Taipei 101, for six years the world's tallest building, features perhaps the most spectacular damper of all—a gold orb that sits on the ninety-first floor and constitutes a prominent part of the building's architecture.

Citicorp scare

The Citicorp Tower, in Midtown Manhattan, opened in 1977 as something of an engineering wonder: its 25,000 tons (22,700 metric tons) of steel rest on four nine-story stilts placed in the middle of the building's sides, allowing it to cantilever over a church on the same block.

Soon after its opening, a Princeton engineering student contacted the structural engineer, William Le Messurier, to ask whether the columns were in the right place to best resist wind loads. In revisiting his model, Le Messurier realized that the welded joints he had specified to hold bracing in place had been replaced, during the construction process, by bolted ones. After doing further wind-tunnel testing, he also realized that his staff had never considered the possibility of nonperpendicular winds (called "quartering winds")—and that he had a big, 59-story problem on his hands.

Le Messurier quietly let Citicorp and its lawyers know about the problem, then came up with a plan to stabilize the building. First, emergency generators were brought in to ensure that the mass damper atop the building—a critical component in fighting wind load—would not lose power. Then Le Messurier developed a plan to weld thick steel plates over the 200 questionable joints. Next he put a form of transponder on individual building components to monitor their stability via a real-time telecom connection. Then he hired weather experts to predict wind forces four times daily. Finally he worked out an elaborate evacuation plan with the Red Cross—just in case the worst came to pass.

The welding work was undertaken in off-hours so as not to alarm the building's tenants. Drywall crews and carpenters worked from 5 to 8 p.m., welders from 8 p.m. to 4 a.m., and cleanup laborers from 4 a.m. on. Working seven days a week, the welding job was only partially complete by the time Hurricane Ella took aim at New York in early September of 1978—but she ultimately headed out to sea and the job was finished in October.

THE SKIN

For the first 75 years of their existence nearly all skyscrapers had windows—traditional windows that opened and closed to allow natural ventilation. But with the rise of the international style in the mid-1900s, traditional windows all but disappeared from commercial office towers (though not from residential ones). Concrete or brick walls punctuated regularly with window frames gave way to sleek, uniform walls of metal or glass—most of which didn't open at all.

These "curtain walls," as the façades of these new buildings came to be known, were defined by a system of metal frames into which panels of metal, stone, or glass could be inserted. This new type of building skin carried no load except its own weight, which it transferred—along with any wind loads pressing on the building—to the building's floor structure through a system of anchors.

The earliest curtain wall was composed of steel frames (known as "mullions") into which glass or metal panels (called "infill") were placed. The seven-story Hallidie Building, located on Sutter Street in San Francisco, is generally acclaimed as the

country's first curtain wall structure. Opened in 1918, it featured a regular grid of mullions that held glass panels in place—three vertical panels per floor. The building was renovated in 1979 but has retained its signature blue and gold coloring for almost 100 years.

But curtain wall technology was not really popularized until three decades later, by which time glass technology had matured significantly. Lever House,

which was designed by Gordon Bunshaft at Skidmore, Owings & Merrill and opened in New York in 1952, featured a skin composed of stainless steel and heat-resistant glass. The windows did not open; to clean them, a window-washing gondola was lowered from tracks along a parapet at the roof of the building. The idea of the building was revolutionary, and its stylish execution gained worldwide recognition.

Hallidie Building, San Francisco (1918)
The Hallidie Building is considered one of the first glass curtain wall buildings.

Lever House, New York (1952)
Lever House is among the earliest "modern" glass curtain wall buildings.

Types of skin

Curtain wall can be made of metal, stone, or glass.

Stone façades are often "hung" from the structure, just like glass curtain wall panels.

Some brick and stone façades must be laid by hand with high-rise scaffolding.

Spandrel glass or panels are often cheaper than vision glass and are selected for thermal performance over transparency.

Vision glass is selected to provide views while maintaining thermal performance.

STONE CURTAIN WALL

GLASS CURTAIN WALL

The opening of Lever House coincided with the development of the "float glass" process patented by the Pilkington Company. The process involved floating molten glass on a bed of molten tin to produce glass sheets of uniform thickness. The process was continuous, allowing a high volume of production of consistently high-quality glass. Most important, the panes it produced were large—with perfectly flat and parallel surfaces.

Commercial manufacturing of float glass began in 1959. Glass soon became the material of choice for skyscrapers around the world—and it remains so today. Of the 53 million tons (48 million metric tons) of float glass produced in 2008, 70 percent was used for buildings; the rest was used for automotive, furniture, and other interior applications. Advances in float forming now allow the production of glass as thin as one one-hundredth of an inch (three tenths of a millimeter) and as thick as an inch (25 millimeters).

Mullions have also changed since the debut of the glass curtain wall building. Today the mullions that glass panels slide into are no longer steel; they are instead made of extruded aluminum, which is made by pouring molten aluminum through a mold or die and then cutting it down to size.

Making Glass

The Pilkington process of making float glass proved the culmination of years of experimentation in glassmaking. Traditionally window glass had been made by blowing large cylinders from molten glass, cutting them into pieces, and then flattening the pieces. Because the cylinders were generally no more than a foot in diameter, the width of the panes that could be cut out of it was limited. To match the size of the glass, windows needed to be divided into relatively small panels by transoms.

In the late nineteenth and early twentieth centuries, a variety of attempts were undertaken to produce a continuous ribbon of flat glass from a molten material—either by putting it through rollers or pulling it upward from a tank—including one by Bessemer, who had successfully pioneered the mass production of steel. None proved easy to commercialize, however, as they required polishing and grinding, which added considerable cost to production.

"Back of house" glass

Not all float glass is created equal. On most skyscrapers, glass that will serve as windows for offices or residences needs to be highly transparent and is generally known as "vision glass." It is usually double-glazed.

In contrast, the glass that covers the "back of house" sections, for example, where the curtain wall meets the floors (often referred to as the "spandrel") or where mechanical floors are located, need not be as transparent. Generally known as "spandrel glass," it may contain a fritting on the external glass or be used in the creation of a "shadow box." Also referred to as "monolithic glass," it is usually one quarter of an inch (6.4 millimeters) thick—half the thickness of vision glass.

The float glass manufacturing process

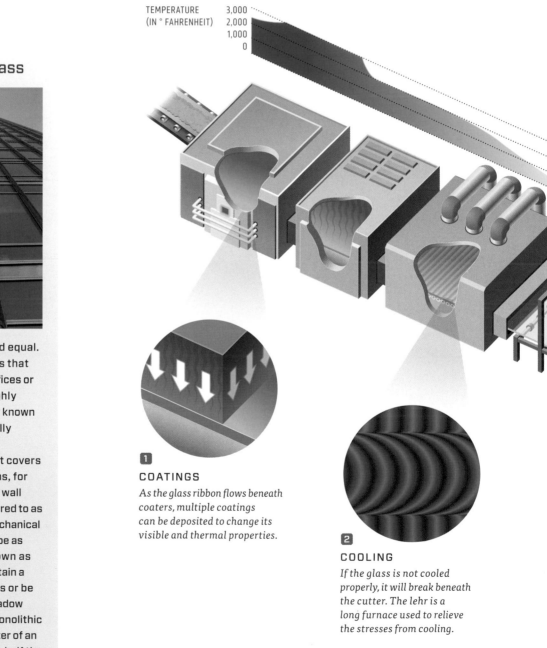

TEMPERATURE (IN ° FAHRENHEIT) 3,000 2,000 1,000 0

1 COATINGS

As the glass ribbon flows beneath coaters, multiple coatings can be deposited to change its visible and thermal properties.

2 COOLING

If the glass is not cooled properly, it will break beneath the cutter. The lehr is a long furnace used to relieve the stresses from cooling.

Where is float glass made?

Around 70 percent of the 53 million-ton (48 million-metric-ton) worldwide float glass market is consumed by windows for buildings. Most of it is manufactured in Europe and Asia, although some is made in the eastern United States and a few plants can be found in South America.

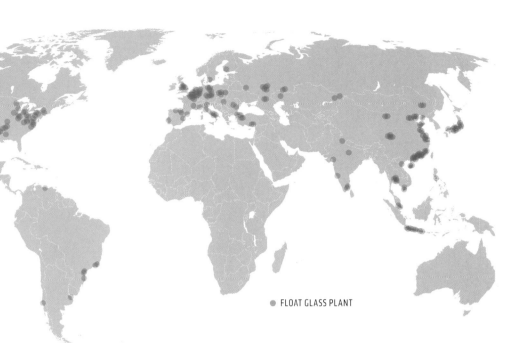

● FLOAT GLASS PLANT

3 DEFECTS

Lasers are used to detect defects invisible to the naked eye—such as bubbles, unmelted sand grains, or tin debris from the float.

PROTECTION

The temperatures needed for the float process are so high that some workers must wear heat-protective clothing.

4 CUTTING

Diamond wheels trim off the stressed edges of the glass and cut the ribbon to a size dictated by computer.

Curtain Wall Systems

There are two main types of curtain wall, which differ primarily in where they are assembled. A "stick system," popularized by a company called Pittsburgh Plate Glass (PPG) in the mid-twentieth century, involves the assembly of curtain wall panels at the building site. The mullions, or curtain wall frames, are installed first, and then panels of glass (or in some cases metal or stone) are inserted into the mullions.

In contrast, a "unit system" (also called a "unitized" system) is composed of wall units that have been previously assembled and glazed in the factory, then shipped to the construction site for erection. The units are generally one-story tall and between five and six feet wide (1.5 to 1.8 meters), with vertical and horizontal connections that allow the individual panels to "snap in" to one another. Because most of the assembly work has been done at the factory, rather than in the field with more expensive construction labor, labor costs associated with unitized systems tend to be lower—offset, not surprisingly, by a higher price for the preassembled curtain wall itself.

Both stick and unitized systems can be assembled from either inside the building's steel frame or from outside it. "Interior glazed" panels are inserted into the frame opening from the inside of a building under construction. While they are somewhat complicated to erect, due to the requirement for obstruction-free interior space for insertion, they are relatively easy to replace subsequently. In contrast, "external glazed" panels are installed from the outside, and therefore require no internal staging area, but do require scaffolding, a crane, or a swing stage to be erected when or if they need to be repaired.

A third system, called "point-supported glass," is now used to create walls that demand maximum transparency—such as lobby façades, large retail windows, or atria roofs. In these cases, glass is attached to the structure by bolted fittings that are connected through holes in the glass. These seemingly "pure glass" walls are designed in varying thicknesses to take account of projected wind loads and, in cases where they serve as ceilings for atria or lobbies, gravity loads as well.

UNITIZED SYSTEM
Unitized panels are fabricated and assembled at the factory. The panels are then taken to the construction site, where they are attached to a building structure. This is the quickest system for installation.

STICK-BUILT SYSTEM
In a stick system mullions (sticks) are fabricated in the factory and installed and glazed in the field. Sticks are placed between floors vertically to support individual components, such as horizontal mullions, glazing, and spandrels.

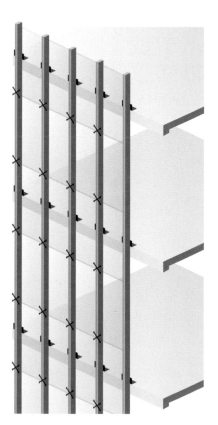

POINT-SUPPORTED SYSTEM
In this system the vertical framing member is a stick, cable, or another custom structure. The glass is supported by a system of four-point brackets and the joints are sealed with silicone.

Installing the glass curtain wall

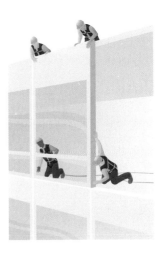

1 *Unitized panels, generally one-story high, arrive on the site fully assembled. Each unitized panel is attached to a crane or davit located on a higher floor.*

2 *The crane lifts the panel into place while workers, generally two below and one or two on top, guide the panel into its assigned place.*

3 *Once in place, the panel is hung from brackets installed on the building and adjusted to be flush with the adjacent panel.*

4 *To complete panel installation, workers snap and then bolt the panel into place and seal its joints with silicone, caulking, or gaskets.*

Diagrid dreaming

When William Randolph Hearst designed his six-story headquarters at New York's Columbus Circle in 1926, he envisioned the area as a magnet for media companies and assumed he would one day build a tower atop his building. The Great Depression put an end to his skyscraper plans, and it would not be until the next century—2003 to be precise—that a tower would begin to rise on his cast-stone façade, now preserved as a city landmark on the corner of West 57th St. and 8th Ave.

The 46-story Hearst Tower, designed by Norman Foster, is notable on many counts—including being one of the first "sustainable" buildings in New York. But it is recognizable above all for the unconventional triangular framing pattern that supports the structure,

known as a "diagrid" (short for diagonal grid). Rather than rising vertically, the building's steel columns are inclined and therefore function as both bracing and column for the building—reducing the amount of steel used in its construction by about 20 percent.

The tower's unitized skin is composed of what appears to be triangular, but is in fact diamond-shaped facets of glass. Each of the four-story facets is 54 feet tall (16.5 meters), with floor-to-floor glazing made up of high-performance, low-emission glass. The curtain wall is accompanied by integral roll-down blinds. Light sensors constantly measure the amount of light penetrating the glass and automatically adjust the amount of electric lighting provided within the building.

Water

Above all else, curtain walls are designed to keep out the elements—primarily water and air. Of the two, the former is more dangerous, as water penetration can rapidly lead to the deterioration of the structure and mold. And yet, even today, all buildings leak: the trick is to "weep out" the water in a controlled and predictable route.

Thanks to advances in technology, less water infiltration occurs today than it did in the early days of skyscrapers. Yet the water that penetrates the curtain walls of modern towers is more problematic than that which penetrated stone or masonry walls, as neither of these materials deteriorated when in contact with moisture. But with the introduction of metal into building frames, and subsequently into building studs and other components, water became a more potent enemy —prompting the development of innovations to ensure its drainage. Today most modern curtain wall systems feature innovations like "weep holes" or "condensation gutters" to allow water to drain from the curtain wall to the exterior of the building.

To prevent water from penetrating the wall in the first place, a variety of sealant systems are used. The simplest rely on "face-sealed" barrier walls, with no provision for removal of water that successfully penetrates the barrier. More sophisticated "water-managed" systems contain drains to provide an exit channel for water that enters the glazing pocket between the panes of the panel.

Even more protection is offered by a third system known as a "pressure-equalized rain screen." The rain screen principle involves creating a pressure-equalized chamber that functions as an air barrier between the outside-exposed face of the wall and the interior one. By eliminating pressure differences between the two, water is prevented from entering the glazing pocket. Because it is highly resistant to air and water infiltration, most unitized systems rely on the rain screen system.

Keeping water out

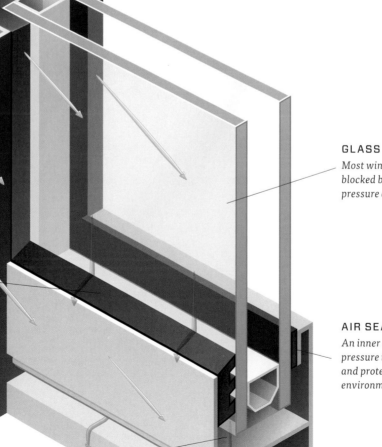

GLASS
Most wind-driven rain is blocked by the glass and pressure chamber baffle.

WATER SEAL
Any water that leaks past outside seals is designed to drain back to the outside.

AIR SEAL
An inner air seal moderates pressure in the chamber and protects the interior environment.

PRESSURE CHAMBER
A pressure chamber with a baffle collects and drains any wind-driven rain.

Noise

The building's skin also acts as a barrier between outside noise and the interior of a commercial or residential tower. Historically, unusually thick or multiple panes of glass (for example, "double glazing") have been used to minimize the penetration of noise into a tall building. The most effective double glazing for purposes of blocking sound is thought to be two panes of glass of differing thicknesses—which for some reason insulate more effectively than two panes of equal thickness.

Where noise presents a serious problem, three panes of glass may be used. "Triple glazing" involves placing an additional pane of glass roughly three inches away from a one-inch insulated unit comprising two panes. The airspace between the double- and single-pane units can absorb a significant amount of additional noise.

Several other technologies may be employed to minimize outside noise infiltration. Sound-attenuating infill, often composed of a composite plastic material, can be placed between panes of glass to further minimize air and noise leakage. A variety of special seals and sealants may also be employed. Gaskets, typically placed between pressure bars (fastened to the outside of mullions to retain the glass panel) and the mullions to minimize heat conductivity, often help to reduce noise as well.

Somewhat ironically, the level of noise penetration in a city rarely decreases with height—in fact it can be noisier at the top of certain skyscrapers than it is at the bottom. At times this can be a function of the proximity of mechanical systems, such as fans, generators, and cooling towers on upper mechanical floors and roofs. But it is more often due to street noise bouncing, or "canyoning," off nearby buildings. Just how loud those street sounds are as they rise is a function of both the proximity of nearby buildings and the materials those buildings are made of: stone and masonry walls absorb considerably more noise than glass.

Canyons of glass

In dense cities, street noise often seems to travel upward with little diminution of volume. Just how pervasive street noise is to a skyscraper dweller dozens of stories above the earth depends not only on the sound absorption properties of the building's skin, but also on the nature of the skin of nearby buildings and the width of adjacent streets. Holding aside changes in traffic volumes, cities have become noisier places than they once were due to the fact that the glass skin of today's buildings absorbs sound far less well than the masonry walls of older buildings.

Light and Heat

Both the metal and glass components of modern curtain walls conduct heat in quantities far greater than earlier stone or masonry building skins did. Aluminum mullions can get particularly hot, so thermal breaks—generally made of polyvinylchloride, or PVC—are regularly placed between the exterior and interior metal layers of the building's skin to lower heat conductivity.

Significant heat can also be carried into a building through glass. Glass is often characterized in terms of "solar gain" and assigned a "solar heat gain coefficient," or SHGC (in Europe, it is referred to as a "G-value"). In addition, glass is also

categorized by how much heat is transferred through it from the outside and is assigned either a U-value or an R-value (its inverse) to represent heat transfer qualities.

Several technologies have been developed to minimize heat transfer and loss. "Insulated glazing" involves reliance on a gas contained in the space between the two sides of double- or even triple-glazed windows and is commonly used in the United States. Less common in the United States, but increasingly found on skyscrapers in Europe and Asia, are "double-skin" curtain walls, which feature an external layer of safety glass complemented by an interior skin of single-

or double-pane glass. Between the two is a vented cavity through which outdoor air either flows naturally or is forced by mechanical means.

Glass is also measured by how much light is able to move through it. Historically, the amount of light transmitted through glass (often referred to as "visible light transmittance," or the "daylight factor") closely tracked the amount of heat transmitted through it—e.g., glass that blocked 70 percent of the solar heat would also block roughly 70 percent of the light. Advances in technology have facilitated the production of glass that today allows more light to pass through than heat.

Types of enclosures

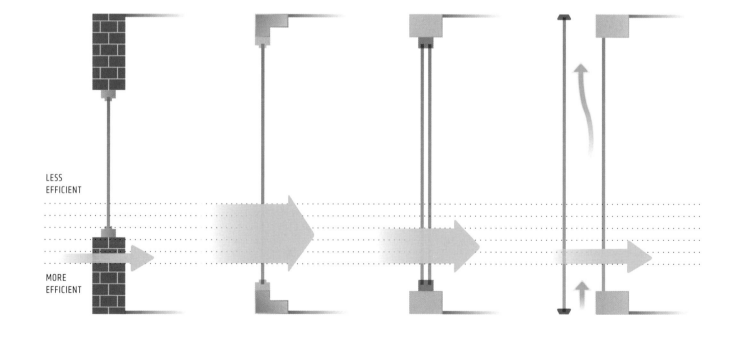

LESS EFFICIENT

MORE EFFICIENT

BRICK VENEER

While not seen in many modern "glass box" skyscrapers, a ratio of approximately 25 to 40 percent window-to-wall area may yield the best compromise between maximizing daylight and controlling thermal performance.

SINGLE PANE

Early glass skyscrapers built in the 1960s and 1970s used only single-pane systems and therefore perform poorly from a thermal perspective.

DOUBLE PANE

Probably most common in high-rise applications today, a double-pane system affords a high degree of transparency as well as some improvement in thermal performance.

INTERACTIVE (VENTILATED)

An interactive or double-skin ventilated system, in which air moves through the space between the glass "skins," is highly transparent and performs similarly to a masonry façade from a thermal perspective.

OVERHANG
Recessed windows or overhangs can reduce solar heat gain by taking the glass out of direct sun exposure.

SHUTTERS
Exterior shutters will also reduce solar heat gain by blocking the sun on exposed faces of the building.

Improving thermal performance

A variety of techniques are available for improving the thermal performance of vision glass in ways that don't obstruct views from the building.

FRITTING
Ceramic fritting, when applied properly to the glass, can reduce solar transmittance without blocking vision, as the eye will read around the frits.

LOW-E COATINGS
Low-emittance coatings, microscopically thin metal or metallic oxide layers deposited on the glazing surface to suppress radiative heat flow, are used widely.

GAS
Filling the void between two panes of glass with inert gases like krypton or argon can further increase overall heat transfer of the glass system.

SHADES
Automatic shading systems, integrated with the curtain wall and controlled by a building management system, can be highly effective.

Killer glass

Both tinted and clear plate glass are invisible to birds. When glass is clear, birds see only what's behind it. When it's reflective, they see only the reflection of sky and trees.

While glass at all heights can pose a hazard to birds, skyscrapers pose a special threat—during the day, because of their height, and at night, because artificial lights attract birds (especially during bad weather). One skyscraper alone was found to have killed over 100 birds in a single day.

To minimize the danger to birds, cities have increasingly adopted light-dimming strategies at night during bird migration seasons in spring and fall. Chicago was the first to implement a voluntary "Lights Out" program in its skyscrapers in 1999; proponents claim it has reduced bird deaths by an estimated 80 percent annually. New York and Toronto, among other cities, have since followed suit and introduced their own light-dimming programs.

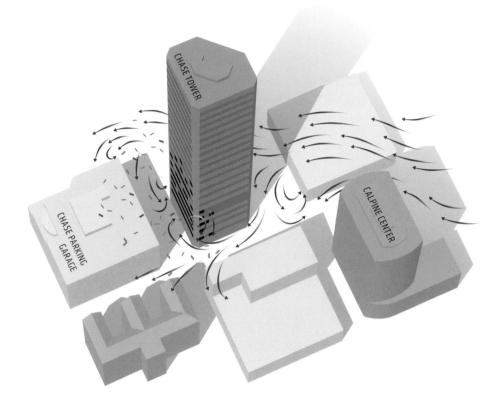

HURRICANE IKE

During Hurricane Ike in September 2008, the 75-story JP Morgan Chase Tower in downtown Houston experienced significant curtain wall damage on the downwind side of the lowest third of the building. The shorter Calpine Center, located upwind of the tower, likely magnified the damage by causing destructive downwind vortices.

Extreme Loads

Standard curtain wall systems contain joints between panels and are built to withstand a certain amount of floor movement (generally less than an inch) without experiencing panel breakage or water leaking in from the outside. To protect against movement larger than that, from wind or an explosion, a range of technologies are available.

One involves the application of a special film upon curtain wall glass and is known as "fragment retention film." It is composed of a polyester laminate, easily colored, and similar to conventional window film, but it is generally thicker and attached with a stronger adhesive. While this film doesn't improve the blast resistance of the glazing system as a whole, it typically prevents glass from shattering inward or becoming airborne—the primary causes of injuries in extreme load conditions.

Other energy-absorbing "catch systems" rely on the capacity of window materials to absorb and dissipate large amount of energy from a blast. In cable-protected window systems, the damaged glass pushes against a cable catch system that in turn acts to deform the window frame itself—capturing energy that would otherwise be transmitted to the inside of the building.

The plywood palace

The 60-story John Hancock Building has been Boston's tallest building for 30 years. Designed by Pei & Cobb, its curtain wall was noteworthy: it used the largest panes of glass possible, with minimal mullions and no spandrels, and featured a bluish-tinted glass that superbly reflected the sky and buildings around it.

But the Hancock proved a difficult building to construct. Its opening was delayed five years, from 1971 to 1976, due to concerns about both its lateral stability and the unexpected settlement of its foundations. To make matters worse, its glass "windows" kept falling out.

On January 20, 1973, in 75-mph (120 km/hr) winds, a total of 65 panels, each weighing 500 pounds (227 kilograms), crashed to the ground. Over the next several months, dozens of others would fall—forcing the city to close the streets around the tower when winds exceeded 45 mph (72 km/hr). The tower was dubbed "the Plywood Palace," a reference to the plywood that replaced glass in over an acre of the building's skin.

Various theories were examined to explain the failure of the curtain wall. The problem was ultimately attributed to movement and heat stresses caused by contraction and expansion of the air between the inner and outer panels of glass. As a result, the building's designers decided to replace all 10,344 double-pane glass panels with single-pane glass—a process that took nearly three years and cost over $5 million.

Blast protection

Of all the components of a building, windows are among the most vulnerable to terrorism. While it may be impractical to design all windows to resist a large-scale explosive attack, it is desirable to limit the potential for injuries due to glass breakage. Some of the most common solutions used to minimize glass-related injuries are laminated annealed glazing, antishatter films, and energy-absorbing catch systems.

A shattering event

DOWNTOWN OKLAHOMA CITY

- ■ BLAST SITE (A. P. MURRAH FEDERAL BUILDING)
- ■ BROKEN GLASS / DOORS

No matter how thick or reinforced it may be, glass is the weakest part of a building's envelope—and the part most responsible for injuries in an explosion or weather-related event. Just how sensitive it is to impact was shown during the bombing of the Oklahoma City Federal Building in 1995. The blast damaged 312 buildings within a 20-block radius. Thirty sustained serious damage (12 have since been torn down), but nearly all were left with broken glass.

To provide additional security and safety, many public buildings rely on "tempered glass" in areas where human impact could occur. Tempering involves heating a sheet of glass and then slowly cooling it with air; the compression that results from the differential in temperature between the inside of the glass and the outside makes it denser than it would be otherwise. This density causes it to break into many small pieces rather than large shards—one reason why it is also used on retail storefronts as well as on side and back windows in automobiles.

PROTECTIVE FILM

Laminated and antishatter films may be placed on glass to hold it together after impact to minimize the amount of glass debris that is launched into the air.

CATCH-BAR SYSTEM

In an energy-absorbing catch system, a cable system is used to absorb significant amounts of energy—keeping the glass debris intact and impeding its flight into occupied spaces.

CONSTRUCTION

f breaking ground for a new skyscraper represents the beginning of construction, it also represents the culmination of a multiyear process of design and documentation. Architects, engineers, and a host of other consultants will have spent months and months conceptualizing the building, refining and developing their designs, doing the detailed construction drawings that need to be completed before bidding, and awarding bids for the individual phases of construction.

While these architects and engineers stay involved in the process, the balance of responsibility during construction shifts to an entirely new cast of characters.

Foremost among these is the construction manager, who is hired by the developer to oversee and coordinate the entire construction process. And coordination is the operative word: up to 75 or so different subcontractors will work on specific aspects of the construction before the project is complete. These "subs," as they are commonly known, represent a wide variety of trades—from plumbers and electricians to steelworkers and carpenters. Each plays a critical role at a particular stage of the building process.

To complicate matters for the construction manager, the process isn't linear. At any one point in time, different trades will be active on different floors of the site. As soon as steel has risen on a

particular floor, concrete flooring will be poured and the curtain wall or enclosures will be raised. Once the floor is protected from the elements, mechanical and electrical work can be undertaken—followed by interior finishes. As a result, the lower floors of any skyscraper job will be well along toward completion while higher floors may still be unenclosed or even unbuilt.

Just keeping track of who's doing what, let alone making sure they are doing it correctly and safely, is a herculean task. It is also a long one. Timing varies greatly from case to case, but a 50-story skyscraper will generally take between three and five years to construct. Add to that 12 to 18 months spent on developing

Construction timeline

The time required to design and build a skyscraper will depend upon the size and complexity of the design, as well as the site location and complexity of site logistics and access. The following timeline, for a one-million square foot skyscraper on a typical site in Manhattan, is representative only.

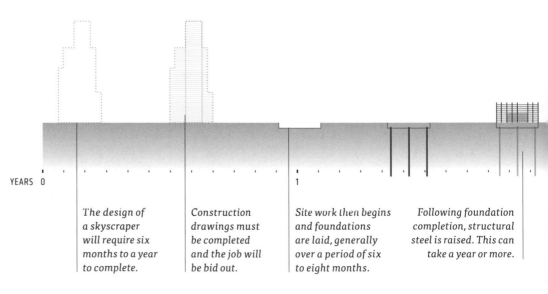

YEARS 0

1

The design of a skyscraper will require six months to a year to complete.

Construction drawings must be completed and the job will be bid out.

Site work then begins and foundations are laid, generally over a period of six to eight months.

Following foundation completion, structural steel is raised. This can take a year or more.

the design and preparing the drawings for construction, and you're talking anywhere from four to six years from start to finish.

In part because of this elongated time-frame, there is an expectation that the building will be constructed as quickly as possible. Time is money to a developer; in the case of a skyscraper, it is a lot of money—typically hundreds of millions of dollars. In most cases the developer has relied on short-term borrowing (often referred to as "construction financing") to cover the costs of construction and is eager to replace this loan with longer-term, cheaper debt when the building is complete. Equally, the developer is likely to have a lease with a tenant or tenants that starts on a particular date—making delays doubly costly.

Men of steel

No ethnic group is more associated with skyscraper construction than the Mohawk Indians, an Iroquois nation located along the St. Lawrence River in upstate New York and Canada. For the Mohawks, the demise of the fur trade in the early nineteenth century cost them their livelihoods. Some turned to farming or timber rafting; others joined traveling circuses or became medicine peddlers. Many remained unemployed and drank heavily.

The arrival of the Dominion Bridge Company to construct a cantilever railroad bridge across the St. Lawrence offered one Mohawk tribe—the Caughnawagas—a way to get its men back to work. In 1886, the right to use reservation land south of the river for a bridge abutment was exchanged for a guarantee to employ the tribe's men during the bridge's construction.

Hired as unskilled laborers to unload wagons, the Indians were repeatedly found walking out on the narrow steel spans projecting high above the river, alongside the highly paid riveters who worked there. Besieged by requests from the Indians to work steel, the company eventually acquiesced and began training them.

The collapse of the unfinished Quebec Bridge in 1907 killed 35 Mohawks but did not dampen the Indians' hunger for steel work. As bridge jobs became less frequent, the gangs shifted to other high-steel work, mainly construction, in Canada. Ultimately they drifted toward New York City—where they would build some of the most famous skyscrapers in the world.

Today there are an estimated 80 Mohawk steelworkers registered to work in Manhattan. Originally from two reservations (the Kahnawake in Quebec and the Akwesasne in Ontario), a small number of them still reside in Brooklyn—just as their fathers and grandfathers did a century ago. Most choose instead to make the 360-mile trip down the New York State Thruway each Monday morning and return home Friday night—an ancestral ritual known as "booming out."

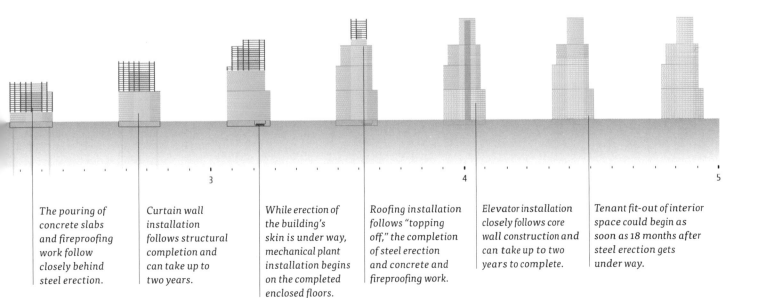

3

4

5

The pouring of concrete slabs and fireproofing work follow closely behind steel erection.

Curtain wall installation follows structural completion and can take up to two years.

While erection of the building's skin is under way, mechanical plant installation begins on the completed enclosed floors.

Roofing installation follows "topping off," the completion of steel erection and concrete and fireproofing work.

Elevator installation closely follows core wall construction and can take up to two years to complete.

Tenant fit-out of interior space could begin as soon as 18 months after steel erection gets under way.

Raising Steel

The first things that rise from a construction site are typically vertical columns of steel. For each of these columns, being lifted and bolted into place marks the end of a long and complicated journey that has generally begun overseas and involved at least three separate journeys—by boat or barge and by rail or road—to arrive at its ultimate destination.

Much of the steel used in the United States is imported from abroad by American fabricators—the middlemen who take sheet steel and turn it into the columns, beams, bolts, and connectors specified by engineers and architects for a particular construction

job. Each component is carefully crafted by the fabricator to match the desired specification, including the number and location of holes, the width and length of the beams, the size of the flanges on its edges, etc. It will then be labeled, numbered, and possibly color coded to ensure that it can be correctly and easily matched to the construction drawings on-site.

Once manufactured, the steel components will be shipped from the fabricator's factory by barge, rail, or truck to a marshaling yard relatively close to the building site. It's the fabricators job to ensure that the

components arrive at the yard in time to be unloaded and catalogued so that they are easily accessible on the day they need to be trucked to the job site.

Steel delivery between the marshaling yard and the job site generally works on a just-in-time principle. Due to space constraints at urban sites, deliveries to the site are typically staggered so that only those pieces to be lifted that day will be delivered early in the morning. Any foul-up at the marshaling yard—e.g., the placement of the wrong set of beams on the delivery truck—can result in expensive and time-consuming delays.

Journey of a steel beam

BARGING UP THE RIVER
Finished columns and beams are loaded onto barges for a three- to four-week journey from Houston to Pittsburgh, via the Mississippi River.

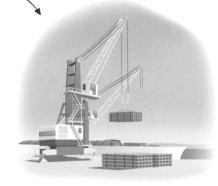

SHIPPING AND FABRICATION
In Antwerp, the steel is loaded on a ship for a four-week journey to Houston, Texas, where it is cut into finished components and connections are added based on measurements from construction documents.

AT THE MARSHALING YARD
The barge contents are offloaded in Pittsburgh onto trucks. One barge will carry the load of approximately 50 tractor-trailer trucks.

ANTWERP PRODUCTION
Raw steel is produced at a steel mill in Luxembourg and trucked to Antwerp, Belgium.

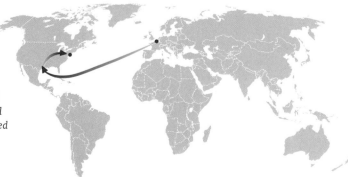

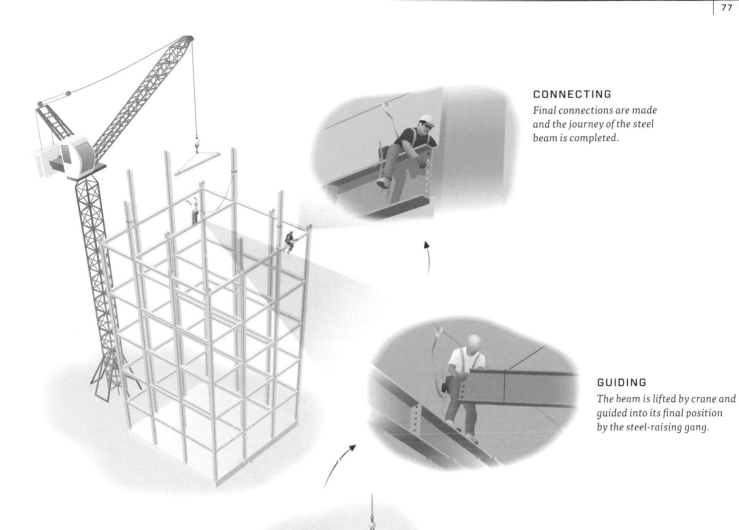

CONNECTING
Final connections are made and the journey of the steel beam is completed.

GUIDING
The beam is lifted by crane and guided into its final position by the steel-raising gang.

SHAKING OUT
The steel needed for the day is trucked to the job site and "shaken out," or laid out in the order and expected position of the day's erection.

TRUCKING TO SITE
The steel is trucked to a marshaling yard in New Jersey, where it will be held until needed at the job site in Manhattan.

Topping off

The completion of the steel skeleton of a skyscraper is often cause for celebration—known as "topping off." Once a quiet Nordic ritual that involved the hoisting of a fir tree or branch to the top of small timber homes to placate tree-dwelling spirits, topping off in its American incarnation has evolved into more of a media event.

As far back as 1912, with the completion of the Woolworth Building, the unfurling of an American flag along a building's highest beam has also been part of the ceremony. Today both fir trees and flags are commonly found in topping off ceremonies; in addition, the last beam up will frequently be painted white and signed by the steelworkers, designers, and project management team before being hoisted into place.

Cranes and Hoists

Once foundations are poured, cranes and hoists become the primary workhorses of a construction site. Cranes are responsible for moving large structural components from the ground to their designated positions hundreds of feet up in the air. Hoists are responsible for lifting the people, equipment, and materials necessary to lock these components into place and to undertake the work that follows frame completion.

While other types of cranes play an important role in foundation work, tower cranes are the ones most associated with skyscraper construction—no doubt because they seem to hang so precariously hundreds of feet above city streets. They are also the most self-sufficient: they take up little or no space on-site, can assemble themselves, and—when their work is done—can take themselves apart almost fully with no outside assistance.

Tower cranes swing in a gentle arc and are erected gradually as the building grows in height. They arrive in pieces—triangulated, lattice components, 10-feet (3.3 meters) square and anywhere from 14 to 20 feet (4.2 to 6 meters) high, ready to be placed one above the other to form the crane's mast as the building grows.

In addition to the mast, tower cranes feature what's known as a "slewing unit" containing the gears and engine, which is located just below the unit's boom. The boom itself reaches out horizontally from the mast, complete with a mobile trolley that runs along its length. The back of the boom contains one or more steel or concrete blocks.

Computerized indicators help prevent the crane operator from lifting more than the crane can handle. Switches monitor both the load on the cable and its distance from the mast to ensure that the crane does not exceed its rated capacity.

TRUCK-MOUNTED AND CRAWLER CRANES

The workhorses of a job site are the tower cranes, but smaller ones, like truck-mounted and crawler cranes, will often be used for lighter lifts on the site.

Guiding a crane

Communication between the crane operator and the ground is often done with radios, but hand signals can also be used.

STOP

RETRACT BOOM

LOWER

TRAVEL

MOVE SLOWLY

The high life

Perhaps not surprisingly, crane operators are among the most highly paid construction workers. But their pay is not based on the height to which they must climb each morning—it's based instead on the length of the boom they operate from their cab high above city streets.

The crane operator is not alone in the sky. He or she shares the rarefied air with an "oiler," an operating engineer responsible for maintaining the moving parts of the crane. The cab is their base in the sky: with amenities ranging from the necessary, such as a heater and portable john, to the nice to have, such as a fridge and a television, both operator and oiler will often stay aloft all day.

The cab is in many ways the nerve center of steel erection. From it the crane operator stays in communication with the "signaler" on the ground via a speaker box. The cab's instrument panel shows the weight of the load and the boom's radius at any given point in time—with warning lights ready to flash if the load approaches the crane's limit. Most cranes also feature an automatic shut off mechanism that is tripped if the lift exceeds those limits.

HOISTS

Made up of one or two cars, or "cages," hoists rely on cable or rack and pinion elevator technology to travel vertically at varying speeds. The two types of hoists found on a construction site bring construction workers and materials, respectively, to whatever floor requires them.

Erected next to the tower, hoists are braced roughly every three floors for stability. Like tower cranes, they are generally leased from companies responsible for erecting and dismantling them. They stay in place until permanent elevators are completed, and are refurbished and relet at the job's completion.

HOIST

EMERGENCY STOP

EXTEND BOOM

RAISE BOOM

LOWER BOOM

USE MAIN HOIST

Tower Cranes

Tower cranes are valued for their self-sufficiency and their ability to do many things besides lifting. For one thing, they can grow in place, thanks to something called a "jumping" or "climbing" frame, which encases the main tower of a crane like a sleeve and relies on a hydraulic lift to raise the top portion of the crane sufficiently to fit in a new section.

Tower cranes can also shut themselves off if they find themselves lifting something too heavy or too far out along the jib; automated devices alert the crane operator when the crane is reaching its capacity. And like certain other skyborne craft, they can refuel themselves in midair in a relatively simple process that relies on gravity.

"Climbing" cranes

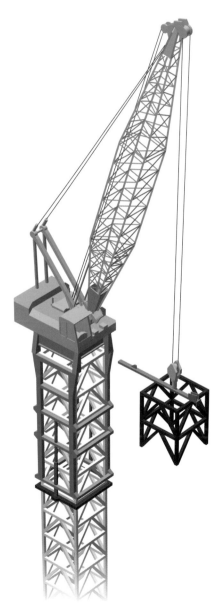

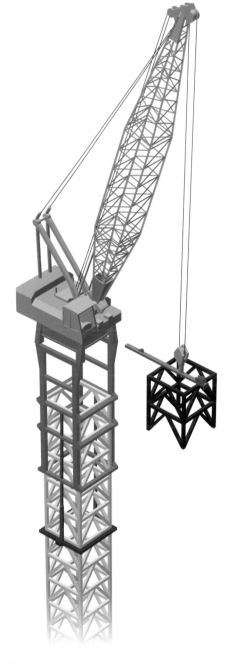

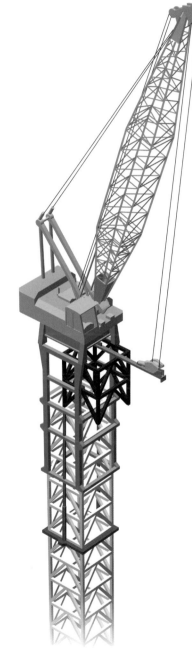

1

The crane operator lifts the new mast section onto a platform on, or suspended from, the climbing frame. The upper portion of the climbing frame is attached below the crane's slewing module.

2

The ram of the hydraulic cylinder is then extended, lifting the top of the crane roughly 14 feet (4.25 meters)— creating an opening large enough to admit the new mast section.

3

After the new tower section is moved into line, it is secured to the existing mast by the erection team. The crane top is then lowered onto the new section of the mast and secured in place by the erection team.

How much it can carry?

Tower cranes work on a simple counterweight principle: the heavy blocks at the back of the boom provide an offset to the weight being lifted at a point along its length. The lifting capacity of the crane varies with the position of the trolley: it can lift greater loads from points closer to, rather than farther from, the mast tower.

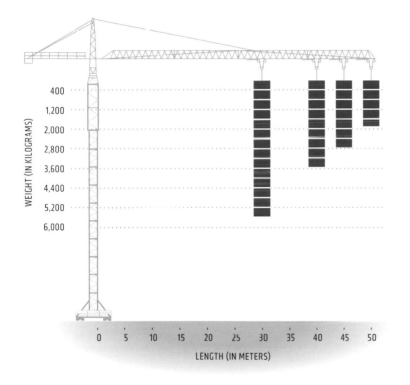

Refueling

Tower cranes are powered by diesel fuel. Roughly once a week the crane operator lifts a tank to cab level and feeds it by gravity into the crane engine's tank. Because they are expensive to rent, tower cranes are expected to work nonstop throughout the day, and generally do—with the exception of bad weather. By law, tower cranes must stop work when winds reach a certain level—usually in the neighborhood of 40 to 45 mph (65 to 80 km/hr).

Crane accidents

Though tower crane accidents happen rarely, they are often fatal. Most often these accidents occur during the jumping or disassembly of a crane. Rarely do they occur when it's fully erected.

Despite the odds, New York experienced two crane accidents within three months of each other in 2008—one involving a fully erected crane and one involving a crane being raised. Both resulted in the deaths of workers on the site; one also killed a tourist passing by.

Like many other cities, New York has for many years required the filing of elaborate plans before a tower crane can be erected. Operators must state where and when it will be built, as well as its radius of movement and the load weight it will carry. Since the tower crane accidents, however, operating rules have become more stringent, and random audits of crane jumps are being conducted by the city's Department of Buildings inspectors.

Concrete

The pouring of concrete marks a critical milestone in the construction of any skyscraper, even those framed in steel. In nearly all cases concrete is used to construct the core of a building as well as its floor slabs. In many residential towers concrete will also be used for the structural frame or support columns.

Concrete is a mixture of cement, water, and a variety of additives. Its consistency is often measure by its "slump," or how fluid it is: the higher the slump, the wetter the concrete. While greater fluidity makes it easier to pump, adding water to concrete can make it less durable and more prone to cracking. High-strength concrete, in contrast, has a lower water-to-cement ratio than average.

While concrete takes on average between 14 and 28 days to gain its full strength, it sets almost immediately. The time that it takes to set varies greatly, depending on climate: in hot weather, it sets more rapidly than in cold. To adjust for climate, a range of admixtures known as "retarders" can be used to slow setting time from a few minutes to several hours; these include simple carbohydrates such as molasses or corn sugar. Other admixtures, called "accelerators," are used to speed up the setting process in cold weather.

The strength of concrete can vary as well. Because concrete is strong in compression and weak in tension, a variety of methods of prestressing it have been developed to mitigate its natural weakness. This generally involves pairing it with steel tendons that are embedded in the concrete at pouring. After three or four days, these tendons are stretched—putting permanent compression force on the concrete and giving it greater strength to offset future tension loads.

Prestressed concrete is commonly used in the construction of high-rises. "Pre-tensioned" concrete is prefabricated and brought to the construction site; "post-tensioned" concrete is poured on-site. Due to their ability to resist deflection, both forms allow floor slabs to be thinner and lighter than they would otherwise be—reducing the building's gravity load and the number of required columns.

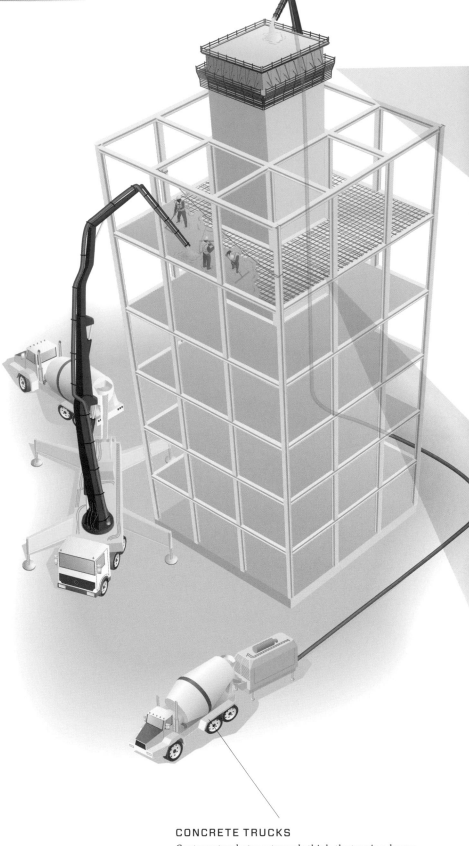

CONCRETE TRUCKS

Contrary to what most people think, the turning drums of concrete trucks are not full of concrete. They leave the factory with a variety of dry materials, which get commingled as the drum turns en route to the construction site. Water is carried in a separate compartment, generally on top of the truck's cab, and is added gradually to ensure the right consistency at the time the concrete needs to be poured on-site.

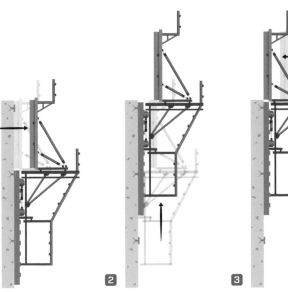

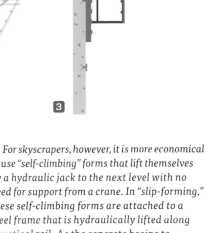

SELF-CLIMBING FORMS

Forms used to shape the concrete core of a building are known as "jump forms" or "climbing forms." They comprise wooden formwork, or molds, of a desired shape that encase the poured concrete until it hardens to a minimum strength, at which point they can be removed and relocated to another floor or place—generally by a crane. Typically the concrete deck and cores, comprising elevator and stair shafts, will be poured individually, and then the formwork will be lifted   with the process repeated over a two- or three-day cycle

For skyscrapers, however, it is more economical to use "self-climbing" forms that lift themselves by a hydraulic jack to the next level with no need for support from a crane. In "slip-forming," these self-climbing forms are attached to a steel frame that is hydraulically lifted along a vertical rail. As the concrete begins to solidify at one level, the climbing form is lifted along the rail and the steel grid is extended—resulting in a single and continuous sheet of vertical concrete.

Pumping high

Concrete trucks are the workhorses of construction sites. Typically they rely on "trailer pumps," steel or rubber hoses attached to outlets on their drums. While these are fine for smaller jobs, for taller skyscrapers a different sort of pump—called a "boom pump"—is required for construction. Its advantage lies in its ability to pump high volumes of concrete, at high pressures, via a remote-controlled robotic arm.

But even traditional boom pumps are not sophisticated enough to meet the altitude needs of today's supertall buildings. For the Burj Khalifa, the world's tallest structure, a new, super-high-pressure pump was developed. Sturdy enough to withstand the enormous forces emanating from the high-strength concrete that was moving through it, the pump required the development of "emergency emptying procedures" so that any batch not placed on a floor within 90 minutes of leaving the mixing chamber could quickly be removed from its length through a combination of compressed air and—in a particularly low-tech touch—a sponge ball.

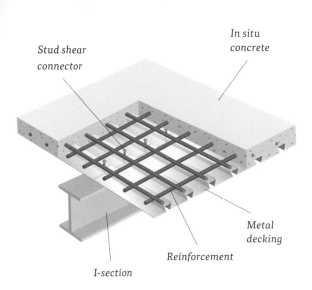

Stud shear connector

In situ concrete

Metal decking

Reinforcement

I-section

ANATOMY OF A FLOOR SLAB

A typical floor section involves panels of corrugated, galvanized metal deck, which have been laid on top of steel beams and welded into place. Somewhere between five to six inches (125 to 150 millimeters) of concrete will then be poured on the decking to complete the floor slab. Alternatively, the floor slab can be composed of precast concrete or can incorporate the decking into a "composite" slab. In constructing floors, deflection caused by the weight of the concrete—particularly in the center of the slab—must be taken into account. To counteract the long-term effects of this deflection, floors can be "precambered," or bent upward, during construction—usually to no more than half the projected deflection.

Inspection

Erecting a skyscraper according to plan is an incredibly complex undertaking—and one impossible to do well without making sure that each step along the way is being done correctly. To ensure that this happens, subcontractors from each trade working on the building are generally required to hire independent inspectors to check their work and report directly to the project manager or superintendent.

Inspection begins at the bottom, with the foundation. Geotechnical engineers ensure that the piles, caissons, or mats are functioning as expected and that no undue settlement has occurred during construction. Footings, where the column loads are transferred to the foundation,

are checked carefully to ensure that connections are robust and have followed the design precisely.

Erection of steel requires particular precision. Surveyors are hired to ensure that all angles are consistent with the drawings and to highlight any deviations from the vertical so that they can be corrected quickly. The surveyor will also make markings on the floor (e.g., to show the center point), so that workers drilling holes or installing partitions are able to follow measurements on the detailed construction drawings more easily.

Both bolts and welding are tested on-site. The integrity of a bolted connection is largely a factor of the tightness of the

bolts. This is tested by a calibrated torque wrench, which measures the amount of tightness applied to a bolt. For welds there are a multitude of nondestructive tests to verify the integrity of a connection. They can range from simple visual inspections or the application of a liquid dye that makes minor surface cracks visible to more complicated tests, such as ultrasonic testing and industrial radiography.

Concrete is also inspected, both on-site and off. Independent laboratories are hired to test the strength of the concrete, roughly 28 days after pouring, to ensure it will perform as expected. In addition, surveyors examine the finished floor slab to ensure that it is smooth and flat.

TESTING CONCRETE

Because concrete behaves so differently based on its ingredients and the weather, testing it is an important part of construction. Concrete is tested both as it is poured, in what is known as a "slump test," and over an extended period of time. On-site, the rate at which concrete "slumps," or shifts its shape, is tested to ensure the appropriate consistency. Simultaneously, a cylindrical batch of this concrete will be put aside for 28 to 56 days to allow it to reach its final strength. That strength is then measured by putting the specimen in a compression-testing machine.

In the United States, the onus has often been on the developer to show that test results, typically produced by private concrete testing labs, meet building code specifications. This system of self-policing has come under fire in a number of American cities, including New York. In 2009, the city found that one concrete testing company had falsified results on over 100 projects, including some of the most prominent civic projects. As a result, the city is setting up its own testing lab and will compare random samples from construction sites to the results submitted by private labs.

The elusive "C of O"

In many countries new buildings cannot be occupied without what is known as a "certificate of occupancy," a document that certifies that the building complies with local building codes. Occasionally, however, skyscrapers have been occupied while still under construction. In Chicago, the Trump Tower (above) began renting out hotel rooms in 2008 while construction continued on its upper floors. That same year, Morgan Stanley moved into the International Commerce Center in Hong Kong while the top third of the building was being constructed. Both were granted "TCOs" or temporary certificates of occupancy—short-term, renewable permits certifying certain floors as habitable.

SLUMP TEST

COMPRESSION TEST

Coming up short

From time to time inspection of a particular aspect of construction will cause a slight delay to the project timetable. But rarely if ever does inspection put a halt to a project, or fundamentally change its scope and program.

Yet in 2008 that is precisely what happened at the Harmon—a 49-story mixed-use hotel and residential tower designed by Foster + Partners as part of the MGM Mirage City Center project in Las Vegas. As the tower reached 20 floors, the project's engineer-of-record noticed problems with the rebar (reinforced concrete) on many of the completed floors. On some floors the rebar was simply spaced wrong; on others the ties that held it together had been cut.

A stop-work order was issued immediately. Despite three months of daily rebar inspection reports that had been submitted by the contractor's consultant, inspectors found major problems on 15 of the 20 completed floors. The building owners were given two options: either move the rebar to the correct location, which would involve complete demolition and rebuilding, or shorten the building to 28 floors—finishing the hotel, but doing away with the 200 condos that were designed to stand above it.

They chose the latter, refunding tens of millions of dollars in deposits to buyers. But in shortening the building they also saved roughly $200 million in construction costs and, perhaps more important, avoided the risk that many of the planned condos would not have sold quickly in a collapsing real estate market. The hotel's opening has been delayed following removal of the contractor.

Checking all the angles

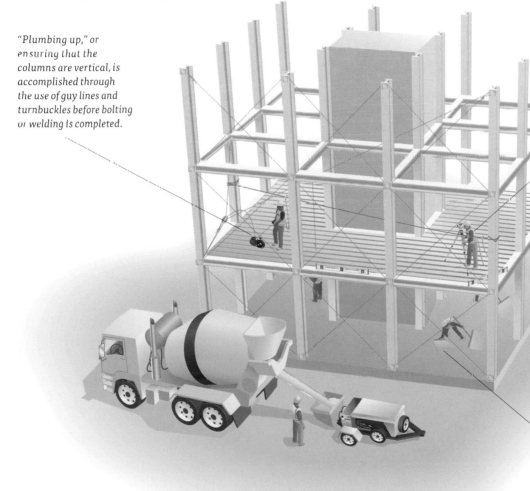

"Plumbing up," or ensuring that the columns are vertical, is accomplished through the use of guy lines and turnbuckles before bolting or welding is completed.

Surveys are used to ensure that columns are properly aligned and straight. Laser levels may be used to ensure the proper flatness of decks.

Manual or vibrating straight edges are used to "strike off" or "screed" the freshly poured concrete, bringing it to proper grade and ensuring a smooth finish.

Who Builds It?

EQUIPMENT OPERATORS

Construction equipment operators run and maintain mechanized equipment at the site, ranging from bulldozers and backhoes to construction hoists.

CONSTRUCTION MANAGERS

Managers oversee the complex orchestration of many trades and are responsible for ensuring the project is completed on time and on budget.

DOCK-BUILDERS

Dock-builders work in the foundation pit, underpinning adjacent buildings and installing shoring or dewatering equipment.

The raising gang

Raising steel is akin to completing a complex jigsaw puzzle—one in which each piece is labeled clearly with its ultimate destination. The center of gravity is also marked on individual steel components, indicating to the signaler and hoisting team on the ground the best place along its length to place the sling.

Roughly half of all erection time is spent hoisting steel; the remainder is spent connecting and leveling the structural members. Once a beam or column is hoisted into its rightful place, bolts are inserted loosely to make a temporary connection. The beam is then "plumbed" to ensure that it is level, before the bolts are tightened permanently by a compressed-air gun. Subsequently, the connection is inspected to ensure that the proper washer is in place and that the bolt has been adjusted to the correct tightness.

SIGNALER

The signaler of the raising gang communicates with the crane operator and hooker-on. The position is usually reserved for older steelworkers.

HOOKER-ON

The hooker-on prepares beams and columns for crane lifting by finding the exact center of each piece of steel. He or she sets the pace of the gang.

TAGLINE PERSON

The tagline person keeps the steel steady during its ascent and safely guides it around obstacles to the connectors at the top.

CONNECTORS

Connectors, working in pairs, shimmy up steel columns and connect beams and columns high above the ground.

CARPENTERS

Carpenters build scaffolding and concrete forms during steel and core construction. Later, during construction of finishes, carpenters will complete sheetrock walls in common areas.

STEAMFITTERS / PLUMBERS

Steamfitters are responsible for installing and welding piping systems that provide heating, ventilation, and cooling. Plumbers install piping systems that connect to washroom fixtures such as toilets and sinks.

TAPERS

Sheetrockers install gypsum wallboard and are followed by tapers and plasterers, who give the spaces a finished look.

BOLTING-UP GANG

The bolting-up gang follows the connectors and temporarily bolts beams and columns in preparation for plumbing and detailing.

CONCRETE WORKERS

Working against time, the concrete workers must keep the concrete pouring from the hoses in a choreographed but continuous process.

ELECTRICIANS

Electricians run electricity cables and high-voltage feeds to connect individual building occupants and fixtures to the grid.

DETAILING GANG

The detailing gang follows the bolting-up and plumbing-up gangs to ensure that the steel is welded or bolted as required.

TIN-KNOCKERS

Tin-knockers, named for their work on large ventilation ducts, install the building's heating, ventilation, and air-conditioning systems.

GLAZIERS

Glaziers install glass curtain wall and form part of a mixed crew that also includes ornamental-iron workers.

Safety

Construction remains a relatively dangerous profession, with somewhere around 50 deaths per 100,000 workers each year in the United States. That makes it less risky than logging or aircraft piloting but more dangerous than many other professions. For that reason safety is paramount on construction sites and is regulated heavily both by the federal Occupational Safety and Health Administration (OSHA) and by localities.

Typically, all workers on a construction site in the United States are required to wear goggles or "safety glasses," as well as a helmet or a "hard hat" to protect them from loose or falling debris. In some cases they are also expected to wear face shields. Due to their exposure to great heat and light, welders must wear greater protection—including heavy leather gloves, protective jackets, and special welding helmets that protect against ultraviolet penetration, which can burn the retina (sometimes referred to as "arc eye").

Depending on the level of protection provided by safety nets, workers must be prepared to "tie off"—or use what are known as "personal fall arrest systems"—to prevent themselves from plummeting downward if they lose their balance on a beam. These systems generally involve wearing a safety harness, which can easily be hooked to a cable running along a beam.

Strict regulations govern "safe" falling distances. In the United States, safety nets or a fully planked floor deck must be located no farther than two floors, or 30 feet (9 meters), from "leading edge" workers. In addition, workers must be able to rely on safety harnesses to prevent falls between 15 and 30 feet (4.5 to 9 meters).

Fall prevention is as important as fall protection. To minimize the number of falls, specific regulations govern the height, location, and appearance of guardrails. They also dictate their strength, as well as the strength of the midrails, screens, or mesh that must form part of the guardrail.

Goggles help to protect against eye injuries resulting from the use of dangerous equipment such as grinders and chipping hammers.

Hard hats protect workers from head injuries resulting from falling debris or tools.

High-visibility vests ensure that workers can be seen.

Gloves protect workers' hands from injury.

Safety body harnesses must be worn by workers who have the potential to fall six feet (1.8 meters) or more.

Steel-toed boots protect the feet and toes from damage resulting from falling materials.

Safety on American construction sites

GUARDRAILS

Guardrails along the edges of floors must be between 39 and 45 inches high (1 meter) and be able to withstand 200 pounds of pressure within two inches (10 millimeters) of the top rail. Midrails, screens, or a mesh must be provided below the guardrail so that no openings larger than 19 inches (0.5 meter) exist; they must be able to withstand 150 pounds of pressure.

SAFETY NETS AND SCREENS

Safety nets must be located within 30 feet (12 meters) of workers and project out between 8 and 13 feet (3 to 5 meters) from the building's edge and have a mesh size no larger than six inches by six inches. To ensure that safety nets are strong enough to take the impact of a falling body, a "drop test" is conducted using a 400-pound (182-kilo) bag of sand roughly two and a half feet in diameter.

Sidewalk sheds must be installed at street level to protect pedestrians walking below from injury due to falling objects at an adjacent construction site.

Unguarded steel reinforcing bars present an impalement hazard. Workers must cover all protruding ends of steel rebar with rebar caps or wooden troughs, or bend the rebar so that the exposed ends are no longer upright.

Safety down under

Some contractors opt to use a self-climbing perimeter-screen system to help ensure worker safety while speeding construction progress. The screen is often designed to protect three to four levels of a construction site at any given point in time.

Depending on the suitability of the shape of the building's perimeter, integrated formwork equipment can be attached to the self-climbing screen. Most popular in Australia, these perimeter screens are now being marketed worldwide.

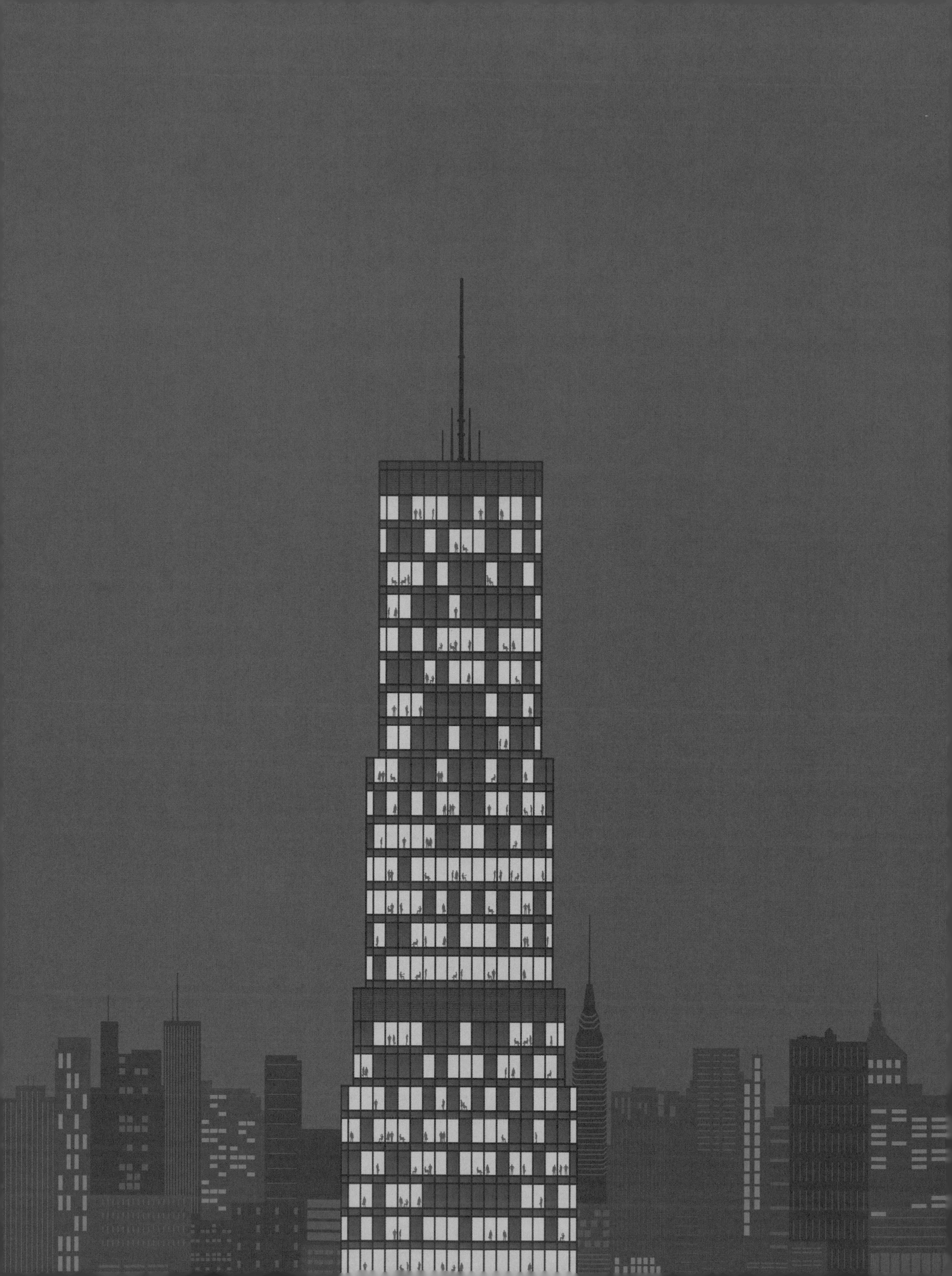

LIVING IN IT

ELEVATORS

While the rise of the skyscraper is often attributed to the invention of the structural-steel frame, its success is in no small part due to the invention of the passenger elevator. For what would be the value of a steel-framed building too tall for people to access by stair? Indeed, the increasing height of these towers over time mirrored the development of elevator technology: commercial buildings got taller and more plentiful as elevators got better and faster.

The earliest elevators predated the first skyscrapers by decades, and were fairly primitive. They carried freight, not people, and were powered by steam—generally relying on what is sometimes referred to as "plunger technology." These freight elevators sat on top of a large piston fitted into a cylinder sunk into the ground; increasing the steam pressure in the cylinder would force the elevator up the hatchway.

Plunger technology was reliable, but it had its limitations: the steel column, or plunger, that raised the car needed to sit in a pit as far below the ground as the building itself was tall. And the technology was slow—suitable for freight, perhaps, but not for office workers in a hurry.

Undoubtedly the biggest breakthrough in the history of elevators came in the early 1850s, when Elisha Otis introduced the "safety elevator," which relied on cables and pulleys for lifting (rather than on plunger technology). His famous demonstration at the Crystal Palace Exhibition in 1854, in which he stood in an open elevator cab and then cut its cables, marked the beginning of the passenger elevator era.

Otis received a patent for his steam-powered elevator in 1861, by which point several of his new inventions had been installed in buildings in New York. But they were cumbersome affairs and moved in ponderous fashion through the few shafts that had been designed to accommodate the new experiment in vertical travel.

The first hydraulically powered elevator was developed by William Hale in Chicago in 1870. The elevator cab moved up and down on top of a long piston inside a cylinder set deep into the ground. Smoother and somewhat faster than steam power, hydraulic power quickly set the design standard for new elevators. However, the hydraulic lift was still slow, and it was not until electric motors could be married to Otis's safe, cable-based technology that cabs would move fast enough to serve buildings taller than six or seven stories.

Initial experiments with electricity in the 1880s and 1890s were not totally successful, as the stepping up and down of electricity led to very sudden elevator starts and stops. But with the introduction of Otis's variable-speed electric motor in 1903, the ride became much smoother. Within a decade, "traction elevators," as the new cable and pulley-based system became known, made their appearance in new office buildings in New York and Chicago.

Over the course of the next several decades, a series of other improvements in elevator performance—including smoother safety brakes and automatic leveling—were commercialized, many by the Otis Company. Safety was also important, leading to innovations such as automatic door opening and closing, load sensors, and motion sensors.

The ingenuity of Otis's invention went well beyond just speed and safety. Because his elevator relied on a counterweight to assist in lifting the cab, it required relatively small amounts of electricity to operate; without the counterweight, elevators would consume on average seven times more electricity than with it. Even today, elevators remain the most energy-efficient aspect of a tall building; they're responsible for no more than about 5 percent of the energy consumed by a fully climate-controlled skyscraper.

Today's elevators come in shapes that would surprise even Otis. Double-deck elevators allow loading and unloading at adjacent floors simultaneously, while twin-shaft systems feature multiple cabs moving within the same shaft. These and others now travel at speeds he could hardly have imagined—limited only by the ability of the human ear to adjust to changes in air pressure.

But perhaps more notable than any of the refinements to the original technology has been the durability of the basic concept itself over the course of the twentieth century. Nearly every one of the thousands of elevators serving urban skyscrapers today still relies on the basic principles of Otis's safety elevator: cables, counterweights, and a catch system to stop the cab's fall in the unlikely event of a cable breakage.

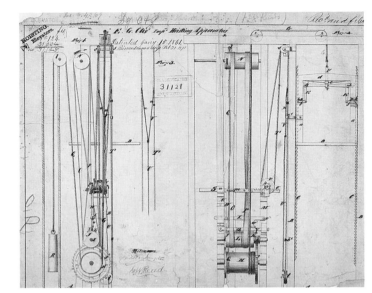

Elisha Otis's elevator patent drawing, January 1861

Elisha Otis at the Crystal Palace Exhibition in London, 1854

How Elevators Work

The basics of traction elevators are the same throughout the world. Woven steel cables are attached to the top of an elevator cab and wrapped in special grooves around a pulley (known as a "drive sheave" or "sheave"). The other end of the cables is attached to a counterweight, weighing roughly the same as a half-loaded cab. The balance between the car and the counterweight means that the motor's only job is to overcome friction.

Both car and counterweight move up and down the shaft on guiderails. These keep the car and the counterweight from swaying, and help stop the car. A variety of safety features exist to prevent the cab from falling through the hoistway, including a "governor" mechanism that senses undue speed and deploys a set of brakes that lock the cab in place along the guiderail. Generally, the sheave, motor, and governor are housed in a machine room located directly above the shaft.

Traction elevators come in two varieties: "geared" and "gearless." In geared elevators, the motor turns a gear that rotates the sheave. The presence of gears makes turning easier and therefore requires a less powerful motor, though it also reduces the speed with which these elevators can travel (350 to 500 feet, or 107 to 152 meters, per minute). In elevators with gearless traction systems, the motor rotates the sheave directly and therefore permits movement at higher speeds (over 500 feet, or 152 meters, per minute).

All elevators share a variety of safety features, including two sets of doors: one on the car itself and one on the floors opening into the shaft. Both are operated by an electric motor hooked up to the elevator's computer. Most elevators also feature load sensors, which tell the elevator computer how full the car is, and motion sensors, which keep the car doors from closing if they detect an obstruction. They also incorporate a computer that registers and processes calls from users.

The main motor is attached to the sheave and turns one way to raise the car and another to lower it.

The sheave is a grooved pulley that rotates to move the wire cables up and down.

The elevator cab has doors that move with the elevator and prevent unintentional falls from the cab into the shaft.

The elevator cab and counterweights ride along guiderails that prevent lateral movement in the shaft.

Wire cables are used to raise and lower the car. There are usually four to eight cables, but one cable can hold a fully loaded car and counterweight.

Each floor has a set of doors that work in tandem with the cab doors and prevent accidental falls into the elevator shaft.

A counterweight, which weighs about 50 percent of the elevator's rated capacity, is attached to the other end of the sheave.

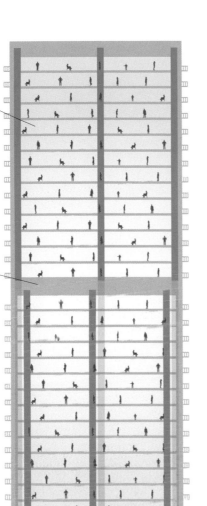

Residential or hotel skyscrapers are designed to accommodate a fraction (one-third to one-eighth) of the number of people on a floor, compared to a commercial building.

Mechanical floors house elevator machine rooms and overruns (in the event of a motor control failure during ascent).

Less densely populated executive offices and law firms place less demand on the elevator system and are typically located on the upper floors of a commercial building.

Floor-to-floor heights are higher in commercial buildings than residential and require the elevator to travel farther to service the same number of floors.

High-intensity uses like trading floors will be located lower in the building to place less demand on the elevator system.

RESIDENTIAL BUILDING
(30 STORIES)

COMMERCIAL BUILDING
(30 STORIES)

Servicing a Tower

Both residential and commercial skyscrapers rely heavily on elevators to move people. But the number, size, and layout of elevators required to serve residential and commercial towers will vary dramatically from one another in almost all cases.

Because natural light need not penetrate the interior of the floors, commercial buildings typically have larger floor plates than residential towers. They are also more densely populated, at least during the day, and exhibit morning, lunchtime, and evening peak travel times. Residential elevator usage is more stable, with smaller numbers moving up and down throughout the day and night.

Elevators in a commercial building will typically be located in a central core area that also serves as home to the wiring and piping that runs vertically through the building. In contrast, elevator shafts in a residential building will be located in different parts of the building so they may directly serve individual apartments.

Elevators in commercial buildings must also travel farther, and therefore faster, as the height of each commercial floor is generally at least 50 percent greater than its residential counterpart. And office workers are impatient, so waiting times must be calculated to fall squarely within the acceptable range.

Elevator Design

Designing a vertical transportation system for a skyscraper is among the most complex tasks developers and their architects face, and involves a series of related decisions. How many elevators are needed? How large should they be? How fast must they travel? How should they be configured—back-to-back or in a straight line? Should a one-stop ride, known as "direct descent," be provided or is an express/local system a better way to move large numbers of people?

Getting these decisions wrong can be very costly to a developer. "Over-elevatoring," or providing more vertical lift capacity than needed, means the unnecessary loss of rentable space. On the other hand, "underelevatoring"

usually results in unacceptable waits and travel times, unhappy tenants, and a bad reputation for the building.

Addressing these design questions requires, as a first step, a series of mathematical calculations. The number of people who will be traveling in and out of the building on a daily basis must be forecast. Generally this is a function of how much square footage is allocated to each employee or resident—a figure that can vary greatly based on the type of work or the sizes of the residences.

A second step is estimating what percentage of those people might move in or out of the building within a peak period known as the "five-minute handling

capacity." This can range from a low of 7 percent for residential buildings to a high of between 20 and 25 percent in office towers, where a large number of employees from one company might share the same working hours.

A number of other criteria factor into the design process, including passenger waiting time, load factors, and total trip time. In each case the developer must decide what level of service it wants to provide to building occupants. In an office building in New York, for example, morning wait times ranging from 20 to 25 seconds are considered good, while those between 30 and 35 seconds are generally considered unacceptable.

Freight elevators

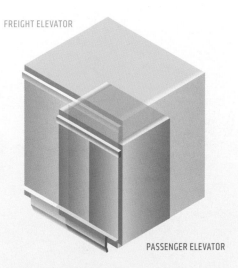

FREIGHT ELEVATOR

PASSENGER ELEVATOR

The largest passenger elevators in skyscrapers are shuttles that ferry people to sky lobbies or observation decks; these can hold weights up to 10,000 to 12,000 pounds (4,500 to 5,400 kilograms) and carry more than 40 passengers. But even smaller skyscrapers will have at least one large freight elevator to move goods and equipment throughout the building. These will typically be several feet deeper and taller than passenger elevators, and will almost always descend directly to an area adjacent to the building's loading dock.

Banking on height

The configuration of elevator banks in a building is largely a function of its height. One bank, generally made up of eight elevators, is usually enough to serve 15 to 20 stories; two banks (a low-rise and a high-rise) are required for buildings of up to 35 stories. Three banks are necessary to reach 40 or 45

stories; four banks will provide service up to 55 or 60 floors. In buildings above 60 floors, a system of express shuttles to one or more sky lobbies, where passengers transfer to local elevators, will generally be needed to minimize the incursion of elevator shafts into the floor plate.

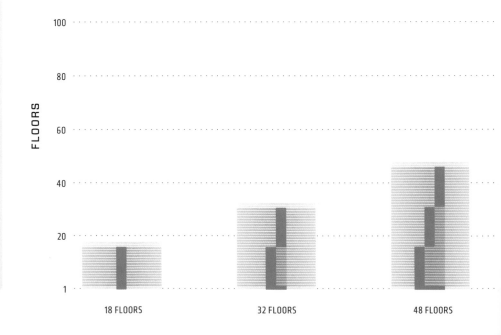

FLOORS

120

100

80

60

40

20

1

18 FLOORS   32 FLOORS   48 FLOORS

Elevator Rush Hour

In elevator lingo, "capacity" generally refers to the maximum number of people a building's elevators can handle during a given five-minute period. In office buildings, a vertical transportation system will typically be designed to handle roughly 12 percent of the building's population in this time. The most complicated peak is at lunchtime, when people are moving both up and down. In residential buildings with less concentrated traffic patterns, elevator systems are generally designed to handle no more than 7 or 8 percent of residents during the morning peak.

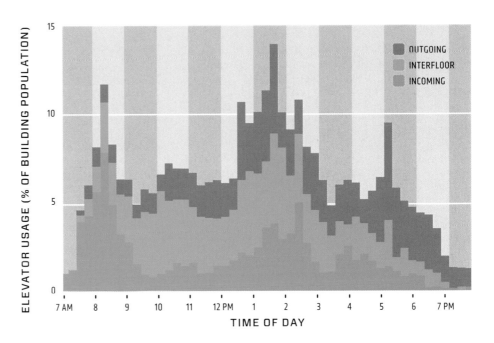

OUTGOING
INTERFLOOR
INCOMING

ELEVATOR USAGE (% OF BUILDING POPULATION)

TIME OF DAY

7 AM 8 9 10 11 12 PM 1 2 3 4 5 6 7 PM

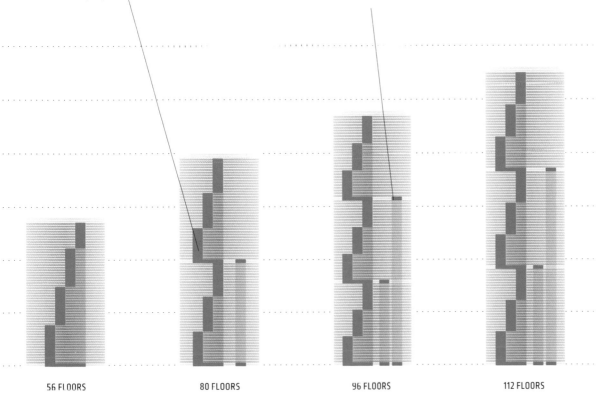

LOCAL ELEVATORS

Passenger elevators come in a variety of sizes and are generally distinguished by the weight they carry. Typically office elevators will be larger than residential ones and will range anywhere from 2,500 to 4,000 pounds (1100 to 1800 kilograms) in carrying capacity—and carry anywhere from 10 to 25 people.

SHUTTLES

Supertall buildings often feature express shuttles to sky lobbies, where passengers can switch to local elevators to access higher floors. This configuration allows the "locals" above the sky lobby to occupy the same shaft as the local elevators serving the lower floors and leaves more rentable space on each floor.

Shuttle elevators made their debut at the World Trade Center in the early 1970s, traveling to sky lobbies on the forty-fourth and the seventy-eighth floors. Each shuttle elevator featured front and rear doors that opened for speedier, unidirectional loading and unloading.

56 FLOORS   80 FLOORS   96 FLOORS   112 FLOORS

New Technologies

While Elisha Otis's cable-based, counterweight-supported safety elevator has remained the backbone of skyscraper elevators around the world, it is continually being improved upon around the margins— mostly in ways that enable developers to move larger numbers of people to their destinations faster, without increasing the footprint of the elevator shafts.

Some of the most interesting innovations are the most disconcerting to the casual elevator user, including advances in elevator-dispatching technology. Historically, elevator brains have assigned cabs to floors based on "estimated time of arrival" dispatching—i.e., the elevator that will get to a floor first in the direction the passenger is going is assigned to collect him or her, regardless of the number of stops it will make before or after.

In new "destination dispatch" systems, the elevator computer relies on a different set of algorithms for grouping elevator users going to the same or adjacent floors. Instead of pushing a button to call an elevator at random, users rely on a keypad or security card to identify their destination at a panel in or near the elevator bank. The computer then indicates a cab assignment. Once aboard the designated elevator, the user will automatically be taken to his or her destination.

Other advances in computerization and elevator "logic" have also served to speed up passenger throughput. By monitoring daily traffic flows, elevator control systems are now able to "learn" where the most efficient place to position cabs might be at a specific time on a given day. For example, passenger flows from one floor of an office to another for a regular 11 a.m. Tuesday meeting will be noted by the system, and one or more cabs will be deployed to "wait" nearby at that time each Tuesday for the expected call from users.

Some advances in passenger-carrying capacity relate to the body rather than the brain of the elevator. Double-deck elevators, which feature an upper and lower cab stacked vertically within one elevator frame, are now in common use. Less common but equally notable are twin-shaft systems; these feature two cabs moving independently within one hoistway.

DESTINATION CONTROL

Traditional dispatch systems are based on minimizing waiting time for passengers without factoring in total trip time. Typically a rider calls an elevator and takes the next available car; the system does not differentiate between a car that subsequently stops at every floor and one that makes a single stop.

Rather than load every waiting person in the car without regard to final destination, the destination-dispatch system aims to minimize total trip time by directing passengers to a specific elevator headed for their floor. The result of grouping passengers in this manner is less crowded cars that make fewer stops and therefore get them to their destination sooner.

TRADITIONAL SYSTEM

4 STOPS 3 STOPS 3 STOPS 3 STOPS

DESTINATION SYSTEM

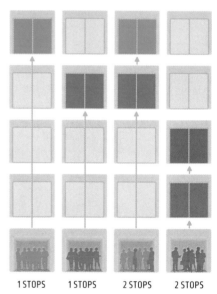

1 STOPS 1 STOPS 2 STOPS 2 STOPS

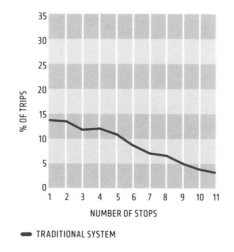

% OF TRIPS

NUMBER OF STOPS

●━ TRADITIONAL SYSTEM

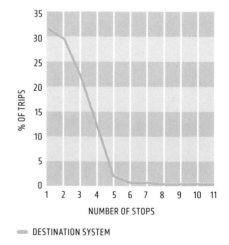

% OF TRIPS

NUMBER OF STOPS

●━ DESTINATION SYSTEM

TWIN SHAFT

The idea of multiple, unattached cabs moving within one shaft debuted in 1931, in a 20-story Pittsburgh office tower. The lower cab served the lower 10 floors and the upper cab the remaining 10. Elevator attendants "driving" these cars relied on blocking signals, similar to those present in subway systems, and the cabs themselves moved only in the same direction. The experiment proved unsuccessful, and twin-shaft technology did not reappear until 2003, when it was featured in a university building in Stuttgart, Germany. In today's iteration the two elevators have their own traction drives, with all moves determined by a destination-dispatch system. Safety features include automatic speed reduction as cars approach one another and emergency brake deployment if minimum safety distances are breached.

DOUBLE DECKER

Double-deck elevators offer the same advantages as double-deck buses and trains—i.e., a doubling of capacity for only a modest amount more fuel. Unlike their transit counterparts, however, the loading and unloading of each deck occurs on different floors rather than through the same doors. Although the technology dates back to an early experiment in New York in the 1930s (which aimed to simultaneously serve the ground- and subway–level entrances of a building), it was not until the late twentieth century that double-deck technology proved itself commercially. It is now employed around the world, including at the Citicorp Center in New York, at First Canadian Place in Toronto, and at the Petronas Towers in Malaysia.

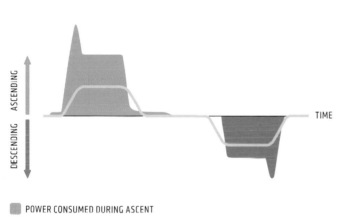

POWER CONSUMED DURING ASCENT

POWER CAPTURED AND USED BY THE BUILDING DURING DESCENT

TYPICAL ELEVATOR SPEED PROFILE

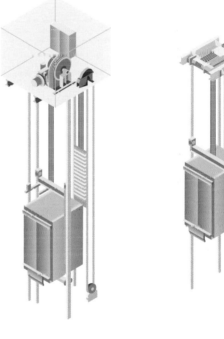

REGEN DRIVES

The "regenerative" technology associated with traction systems can deliver significant energy savings by converting normally wasted energy into electricity. As the cab descends, the elevator's motor acts as a generator and pumps current back into the building's power supply—rather than just dissipating it as waste heat.

MACHINE ROOM–LESS

Traditionally the engine, governor, and controller mechanism are located in a machine room located above the elevator shaft. In "machine room–less" elevators they are located in a structure housed within the hoistway itself— avoiding the construction costs and space requirements of a typical machine room. This is possible due to a new form of flat, polyurethane-coated steel belt, which is significantly thinner than steel cables and permits a smaller, flatter sheave mechanism. Rarely used on buildings above 30 stories, it can prove cost-effective in the design of smaller skyscrapers.

An overspeed safety governor stops the elevator by engaging flyweights and locking the sheave once it reaches an unsafe speed.

Upon loss of power, an electromagnetic brake will engage the guiderails and prevent further elevator motion.

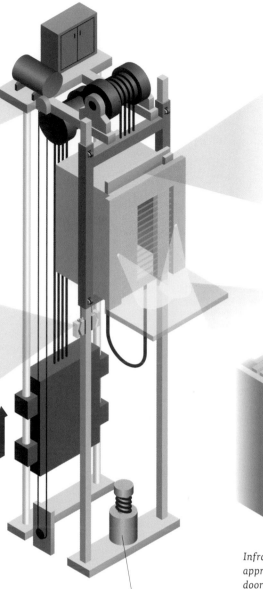

As a last resort a hydraulic buffer is at the bottom of the elevator to stop the car with a deceleration rate of less than one "g."

Door interlocks will prevent an elevator from leaving a floor if the doors are not fully closed.

Infrared beams will detect passengers approaching an elevator and prevent doors from closing and causing injury.

Elevator Safety

As far as urban transportation goes, travel by elevator is far safer than traveling by train, car, or bus. The roughly two dozen deaths a year associated with elevators generally involve maintenance workers falling into an open shaft or getting entangled in elevator machinery. (A rare documented case of an elevator free fall occurred in 1945, when an Army B-25 bomber hit the Empire State Building in fog and its landing gear fell down an elevator shaft, snapping the elevator cables. The cab fell over 75 stories, but the elevator operator survived the fall: compressed air under the cab slowed its fall, and the severed cables, hanging from beneath the car, piled up in the pit and acted as a coiled spring.)

Most skyscraper dwellers are wholly unaware of the numerous safety mechanisms designed into modern traction elevators. Foremost among these mechanisms is the governor, which senses the speed of travel and can automatically bring the machine to a stop by engaging a locking mechanism along the guiderail. It is by no means the only braking mechanism: electromagnetic brakes, kept in an open position for travel, automatically engage when the car loses power for any reason. Likewise, an automatic braking system near the top and bottom of the shaft will be triggered if the car moves too far in either direction.

Other safety features include sensors, which detect people or objects in the doorway and stop or reverse door closures; interlocks and door restraints, which prevent the elevator from leaving a landing if both sets of doors are not closed or from opening when it is between floors; emergency lighting and telecom systems; and a buffer system at the base of the hoistway pit, which works like a cushion to soften any unlikely landing.

When Power Fails

When power fails in a skyscraper, the safety brake on individual elevators is automatically triggered. Mounted between the motor and drive sheave, the brake operates like a spring and is at all times kept in a retracted, or open, position by electricity. A power failure will thus cause it to extend and engage immediately.

In most cases the loss of electric power also kick-starts an emergency generator—usually within 60 seconds of the power failure. Once the generator is triggered, emergency lighting comes on. Generator power then brings the building's cars, one by one, back to the lobby, where their doors are opened, lights are dimmed, and power supply is shut down.

If for some reason, the elevator doesn't move immediately to the lobby, an "alarm," or "help," button within the cab is available to call for assistance. A telecom connection to the building's operations center is also provided. In some cases this connection is two-way; in others it serves only as a channel for incoming calls and instructions.

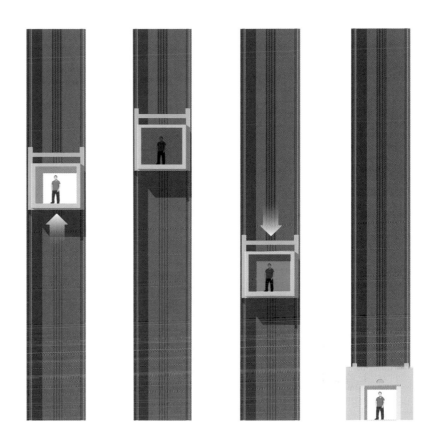

Elevator rescue

Most elevator rescues are made by technicians who fix a computer glitch and get a stuck car moving again. However, when a car is stuck due to a prolonged power failure or some other electromechanical problem, firefighters or other rescue personnel may be called in to free stuck passengers.

Similar to the "jaws of life" used to free passengers from automobiles in car accidents, a pneumatic device may be used by rescue personnel to free a stuck door.

Almost all elevators have a door at the top that rescue personnel can use to gain access via the elevator shaft. Untrained personnel should never use this door to try to escape.

Elevator Speed

Although the basics of traction-elevator technology have remained largely the same for the last century, elevator speeds have increased greatly since Otis's day. Today's high-speed elevators can and do travel as fast or faster than airplanes climbing or descending—and suffer many of the same human limitations.

The world's tallest buildings, many of them in Asia, are home to the world's fastest elevators. This tie between vertical speed and skyscrapers dates back to the Woolworth Building, which boasted the world's fastest elevators (traveling 650 feet or 200 meters per minute) when it opened as the world's tallest building in New York in 1913. Less than 20 years later, the Empire State Building showcased even faster elevators—at 900 feet (275 meters) per minute. The John Hancock Tower in Chicago set the record for the supertall buildings of the 1970s, traveling at over 2,000 feet (609 meters) per minute.

Today's high-speed elevators travel significantly faster than ever before. The Yokohama Landmark Tower debuted in Tokyo in 1993 with elevators reaching 2,300 feet (700 meters) per minute, or over 26 miles (42 km) per hour. A decade later its elevators were surpassed by those of Taipei 101—two of which reach the incredible speed of over 3,000 feet (914 meters) per minute, or 34 miles (54 km) per hour.

The limits on elevator speed are not technical—they are human. Riders can be affected by many aspects of speed, including the jerkiness of stopping and starting, the noise and occasional horizontal sway associated with hoistway travel, and—perhaps most important in the case of skyscrapers—the impact of air pressure changes.

Air pressure changes are particularly limiting. Similar to air travel, changes in air pressure between the inner ear and the external environment are less troubling on ascent than they are on descent—when the eustachian tubes that equalize pressure between the two can fail to open and painful "ear block" can result. In deference to this reality (and after a legal suit against the Sears Tower for eardrum damage), all high-speed elevators travel up significantly faster than they do down.

New technologies can reduce the impact of high-speed elevator travel on passengers. Suppressers and insulation minimize vibration inside the cab. Aerodynamically shaped cars reduce the noise associated with high-speed movement through the hoistway. Double-paneled construction can make cars airtight and prevent whistling. And in the most extreme cases, including the elevators at Taipei 101, the cabin can be gradually pressurized as it descends.

→ AIRFLOW

NOISE

Traveling at maximum speeds nearing 40 miles per hour (64 km/hr), elevator cabs and components are designed using computer simulations and aerodynamic principles similar to automobiles and aircraft to minimize noise for passengers as well as building occupants near the shafts.

The world's fastest elevators

Over time, maximum elevator speeds have tracked closely with skyscraper height records as building designers found efficient ways to move larger numbers of people higher in the sky.

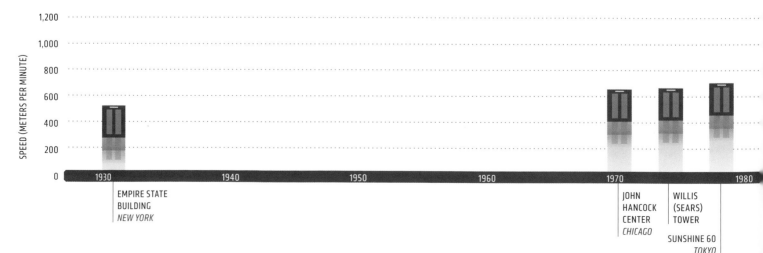

SPEED (METERS PER MINUTE)

1,200
1,000
800
600
400
200
0

1930   1940   1950   1960   1970   1980

EMPIRE STATE
BUILDING
NEW YORK

JOHN
HANCOCK
CENTER
CHICAGO

WILLIS
(SEARS)
TOWER

SUNSHINE 60
TOKYO

Aerodynamic parts

The elevator cabs in Taipei 101 and Burj Khalifa are designed as streamlined capsules, which minimizes friction and flow noise.

Counterweights may also be streamlined to cut down on flow noise heard by building occupants as these weights "fly" through the shaftway.

Asia's fastest elevator

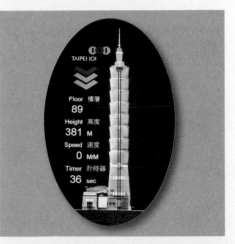

Until recently Asia was home to the world's fastest elevators. Traveling express from the ground to the observation deck at Taipei 101, two high-speed elevators reach over 3,000 foot (915 meters) per minute, or 34 mph (54 km/hr), on the trip up—a full 30 percent faster than any other elevator in the world at its opening—and roughly 1,900 feet (580 meters) per minute, or 22 mph (35 km/hr), on the trip down. Only the observation deck elevators at the Burj Khalifa, which opened in Dubai in January 2010, are designed to ascend faster— at 3,500 feet (1,076 meters) per minute, or 40 mph (64 km/hr).

Both the cars and the counterweights at Taipei 101 are shaped like bullets, to reduce drag. To reduce noise the elevators feature acoustic tiles and sound insulation shrouds. Because conventional bronze safety shoes would melt under the high braking temperatures, ceramic braking shoes were used instead. In addition, a sophisticated system of vibration suppression was designed specifically for the building.

But the most notable feature of Taipei 101's express elevators is its air-pressurization system, which is designed to protect the ears of observation deck visitors en route back down to the ground. It relies on suction and discharge blowers to adjust the atmospheric pressure in the elevator car, beginning as soon as the doors close at the top and continuing at a uniform pace as the car descends.

In addition to its express elevators, the skyscraper has another 59 elevators, including 34 double-deckers and 25 individual large-capacity elevators.

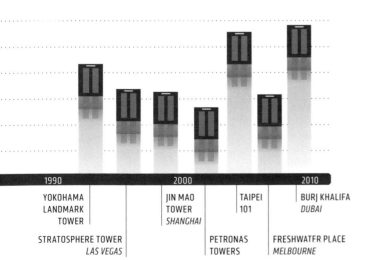

1990		2000			2010	
YOKOHAMA LANDMARK TOWER		JIN MAO TOWER *SHANGHAI*		TAIPEI 101	BURJ KHALIFA *DUBAI*	
STRATOSPHERE TOWER *LAS VEGAS*			PETRONAS TOWERS		FRESHWATER PLACE *MELBOURNE*	

POWER, AIR, AND WATER

By the beginning of the twentieth century, steel framing of buildings had supplanted load-bearing walls, and safety elevators had become an accepted means of moving people vertically in tall buildings. But a host of things would be necessary to enable space high up in the new towers to fully accommodate human life—specifically, light, air, and water.

Advances in lighting technology were particularly important to the evolution of the skyscraper. Gas lighting could never have migrated successfully to tall buildings: it was labor-intensive, messy, and far too dangerous. But Edison's invention of incandescent lighting, or as we know it, the electric bulb, made lighting office buildings both simple and safe. Not only could power be provided remotely, at a generating station off the premises, but each light could be turned on and off separately by the end user rather than being tied to one central switch.

Incandescent lighting was a major step forward for office workers, but it demanded careful design on the part of skyscraper architects and engineers. Because the light desk-top lamps gave off was localized, buildings had to be designed to maximize sunlight—and even when they did, office floor depths would only rarely extend beyond 20 or 25 feet (6 to 8 meters).

New York and Chicago followed different paths in the pursuit of natural light. Chicago, with its larger floor plates, favored atria and courtyards to bring sunlight into a building. New York developed "alphabet" architecture on its smaller plots—with buildings shaped in variations of the letters H, E, and I, among others, to bring outside light into buildings sitting on narrower, deeper lots.

The development of fluorescent lighting in the 1940s freed designers from such limitations. The brighter light, covering a wider area, meant that offices could be farther away from windows, permitting more regularly shaped, and bigger, floor plates. Today fluorescent lighting remains the choice in most commercial towers—and because of its energy efficiency compared with traditional incandescent lighting, it is increasingly finding a home in residential towers as well.

Developments in lighting were accompanied by developments in

Building inventions

The second half of the nineteenth century saw a series of inventions relating to power, light, and water that made life in the sky possible.

| 1840 | 1850 | 1860 | 187 |

FLUSH TOILET   ELEVATOR

mechanical ventilation—specifically, heating and cooling. As with electric power, the ability to provide heating centrally within a building—as opposed to having each office user fend for him- or herself— proved critical to housing large numbers of people in one place.

The earliest central heating systems relied on forced air heated either by wood or coal fires. By the turn of the century, however, radiators—which relied on hot water or steam circulating from a central boiler or steam system—were being designed into building systems. By the time the Empire State Building was constructed in the late 1920s, radiator technology was well advanced: the tower had 7,000 radiators served by four separate riser systems from the basement.

Central air-conditioning would not come to skyscrapers until somewhat later, though individual air conditioners could be found in both residential and commercial buildings as early as the late 1920s. It too proved important to skyscraper evolution: the fully enclosed glass curtain wall buildings that epitomized the international style of the mid-twentieth century would never have worked without it.

One final innovation of the Industrial Revolution period was also critical to life in the sky: running water. Without the development of both domestic plumbing (i.e., toilets and sinks) and municipal sewers (to drain them into), it would have been both unpleasant and unsafe to design buildings with thousands of people living or working on top of one another. But by the late nineteenth century internal plumbing appliances had become standard in new buildings, and municipal sewage disposal—at least in bigger cities like New York and Chicago—had become a reality.

Prior to World War II, offices were dimly lit by incandescent lights and relied on cross ventilation for cooling on warm summer days.

After World War II, the commercialization of air-conditioning and fluorescent lights led to artificially lit spaces and the demise of natural ventilation in offices.

| 1880 | 1890 | 1900 | 1910 | 1920 | 1930 | 1940 | 1950 | 1960 |

INCANDESCENT LIGHTS   ELECTRIC MOTORS (VENTILATION)   RADIATOR   AIR CONDITIONING   FLUORESCENT LIGHTS

Mechanical Floors

Today the heart of any tall building is its mechanical floor. Light, water, and air all emanate from it—although the machines that provide the first two of these three are really just glorified connections to municipal power and water systems. Unlike smaller buildings, which often feature just one (usually the basement), skyscrapers generally feature multiple mechanical floors—roughly one for up to every 30 floors in height.

Mechanical floors in skyscrapers can be located anywhere. Often they are spread out at different levels, to more easily serve floors below and above and to assist in managing water pressure within the building. Occasionally this distribution dovetails with the structural needs of the building—e.g., unrentable space housing outriggers (which tie the frame to the core) may also house mechanical equipment.

Frequently the top floor of a skyscraper is used as a mechanical floor, and it is referred to as a "mechanical penthouse." It contains the machine rooms for the tallest elevators and also provides a home to any telecom or window-washing equipment that needs to be located on the roof. In the largest of buildings, mechanical rooms may be located strategically above upper-floor lobbies, to separate elevator shafts stacked on top of one another.

Mechanical floors contain a wide variety of equipment supporting the building's systems: chillers for air-conditioning, water pumps and tanks for plumbing, boilers and pumps for heat, and numerous types of telecom and electrical equipment. Given the amount of heavy machinery at work, they require special ventilation, so rather than being enclosed in a glass curtain wall, their perimeter will often be wrapped in external vents or louvers.

Life after CAD

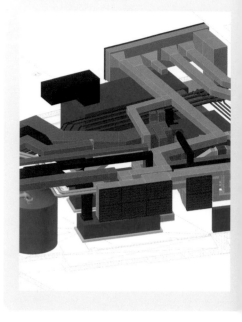

VENTILATION
STAIRS / ELEVATOR SHAFTS
PLUMBING / ELECTRICAL / TELECOM RISERS
ELECTRICAL SWITCHBOARD

The mechanical room shown here contains a large amount of ventilation equipment. Larger rooms may contain generators, pumps, tanks, and ducts for the electrical, heating, cooling, and plumbing needs of the building.

BIM, or building information modeling, is transforming the way skyscrapers are built. It involves the creation of a digital 3-D image containing data about both the physical and operational characteristics of a building under design that can be used by architects, engineers, and construction managers to make sure the plans they are developing for the building are consistent with those being made by others.

Prior to BIM's debut, architects and engineers relied on CAD, or computer-aided design, for a graphic representation of what a particular element of a building—e.g., a beam—would look like. But CAD was "dumb" in the sense that it could not convey nongraphic information. In a BIM visualization, however, the graphic representation of a beam will contain information about the material the beam is made of, its dimensions, the number and location of its bolts, etc.

Initially popular with architects, the use of building information modeling is now growing rapidly among engineers and contractors. Contractors have found that BIM helps avoid errors and conflicts in construction, making it easier for them to deliver projects on time and within budget.

Risers

Both air and water circulate continuously, to every floor in a high-rise building. They move to and from mechanical rooms, where outside air is conditioned (heated or cooled) and filtered, through risers.

- ■ HOT AIR AND WATER
- ■ COLD AIR AND WATER
- □ RETURN AIR AND EXHAUST

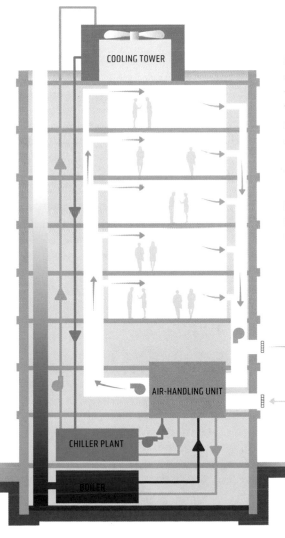

COOLING TOWER

AIR-HANDLING UNIT

CHILLER PLANT

BOILER

Can you see its heart?

To the untrained eye, skyscraper floors—particularly in glass curtain wall buildings—have a certain uniformity. But look closer, and it becomes relatively easy to pick out the mechanical floors on many modern skyscrapers.

JIN MAO
There are only two separate mechanical floor locations—one about two-thirds of the way up and another, larger space at the penthouse level.

WILLIS (SEARS) TOWER
There are five multistory mechanical floors, which correspond to many of the setback levels in the tower.

TAIPEI 101
There are 11 official mechanical floors (of one or more stories) that correspond to the groupings in the office section of the building. Floors 92 to 100 contain communications equipment.

Power

Skyscrapers are consumers of huge amounts of power—not surprising, given the large numbers of people they house. But their actual consumption patterns can vary greatly by place and type of use, with office buildings—because of their larger areas of lit floor space and greater air-handling demands—consuming roughly three to five times the amount of electricity of residential buildings.

Electricity generally arrives at both commercial and residential buildings from the same source. Power produced at a generating station travels at high voltage over a transmission line to an area substation, where its voltage is reduced and sent to a local substation near the end user. There its voltage is stepped down again, to the level at which it can be distributed to businesses and homes in the area.

In some cases buildings meet some of their own power needs on-site. Generally referred to as "distributed generation" in the power industry, and often known as "cogeneration" in the real estate business, there may be a small electricity-producing facility dedicated to supporting part of the building's load. The machinery must conform precisely to local utility standards or the facility will not be able to plug in properly to the broader electrical grid supporting the building.

The reliability of the power supply varies from country to country and city to city. In the most sophisticated of cities power provided to large buildings will rarely fail because of a problem with a localized component. Distribution networks are interconnected, with multiple "grid feeders," or channels, able to carry the load from the supply source to a large end user and support each other should one fail. Most large urban electricity-distribution networks contain at least some level of redundancy.

Nevertheless, most larger skyscrapers are designed with the ability to produce emergency power in the case of a wider power failure. Most frequently, this system relies on diesel-driven generators. The transfer from normal operations to emergency power is intended to be seamless: an automatic transfer switch connects the two, and any power failure is designed to trigger a battery-operated mechanism that starts the generator.

Emergency power is tied into a number of important building systems, with elevators being perhaps the most obvious. Emergency power will also be tied into fire alarm systems, emergency lighting, and exit signs to ensure that occupants can leave a building safely. It will often further support the operation of the firefighting system itself—including electric motor pumps for fire sprinklers, smoke evacuation fans, and dampers, which seal off ventilation zones to prevent the spread of smoke and fire.

Power to the people

Power plants supply energy to large skyscrapers at the same time as they supply power to much smaller end users. However, in high-density or high-load areas, tall buildings will often be connected via a "spot-network" system of distribution—to provide additional redundancy and reliability to large numbers of users that are located in one place.

— 120 TO 480 V
— 13.8 TO 22.0 KV
— 138 KV

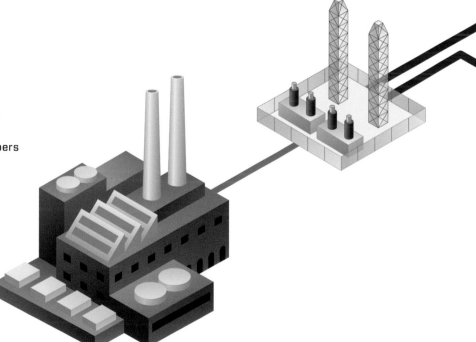

Electricity is generated at 13.8 kV to 22.0 kV and stepped up to 110 kV to reduce the amount of energy lost during transmission. It is then stepped down again, at area and local substations, to lower voltage for delivery to end users.

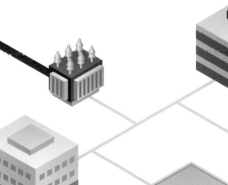

A spot-network system, typically found in certain high-load areas, serves customers at a single location—such as a high-rise office building—through two or more paths. This provides greater reliability in the event of a line or transformer failure en route. Once energy arrives at the building, a switch gear starts the distribution of power within the building—sending it to machine room floors a well as to risers headed for tenant floors.

Energy for commerce

Surprisingly, most energy use in an average office building does not come from computers or other office equipment. Instead, air-conditioning and lighting comprise a full 70 percent of a building's energy consumption; office equipment consumes an additional 20 percent. The balance of the energy is consumed by ancillary uses, such as cooking, refrigeration, water heating, and other building loads.

OTHER	12.4%
WATER HEATING	2.5%
OFFICE EQUIPMENT	6.3%
SPACE HEATING	4.7%
REGRIGERATION	10.7%
VENTILATION	12.3%
COOLING	13.5%
LIGHTING	37.6%

BREAKDOWN OF ENERGY USE IN AN
AVERAGE COMMERCIAL BUILDING

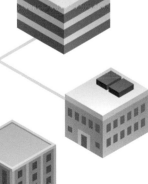

A radial network leaves the substation and passes through the network area without supplying other end users. A single utility line runs from a transformer to the end user.

Ventilation

Although office workers in skyscrapers are rarely aware of it, the air they breathe is being changed constantly. If it weren't, their office would be a much less comfortable—and a much more dangerous—place to work.

The process of replacing air in a building with fresh, outside air is generally referred to as "ventilation." Just as they are in a home, ventilation systems are designed to control temperature and remove odors, bacteria, and moisture from respiration. But while air circulation within a home most frequently relies on natural ventilation, within an enclosed tower it generally relies primarily on mechanical, or "forced," ventilation.

Some movement of air within a skyscraper will occur naturally, due to pressure differences caused by wind or the chimneylike "stack effect." But it is not enough to move sufficient quantities of air through the building, Particularly in situations where the glass curtain wall cannot be penetrated or opened, getting both airflow and air quality right is critical to human health.

Ventilation of floors has historically been accomplished through a complex web of ductwork in the ceiling plenums, or spaces between the ceiling and floor slab above. In addition to designated supply and return channels, ceiling ventilation systems include a series of diffusers located in strategic locations above the office floor. These serve to mix the new "supply air" with the existing room air to maintain a consistent temperature throughout the room and evenly distribute the conditioned air.

While ceiling ventilation remains the predominant way air is brought to commercial towers, underfloor air-distribution systems are gaining popularity. These systems are designed to provide clean conditioned air in the occupied zone, or the lowest six feet of airspace. Due to natural buoyancy, the hotter dirty air rises to a higher level, where it can be exhausted out of the space mechanically. Although the initial cost of underfloor systems is high, it uses less energy, and the air entering the occupied zone of the space tends to be much cleaner.

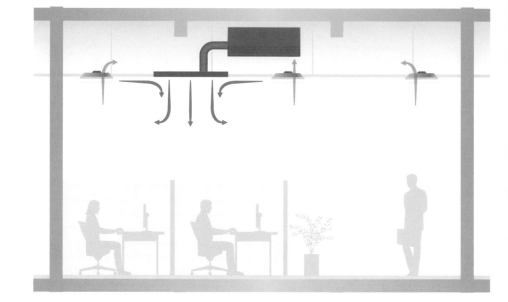

TRADITIONAL CEILING VENTILATION

In traditional systems fresh air is supplied to an office floor from one or more openings in the ceiling while existing room air is extracted from another. The system is designed to supply sufficient new air to condition the entire space. Airflow to different zones will be based on location (e.g., perimeter versus interior) and can be controlled locally by thermostats.

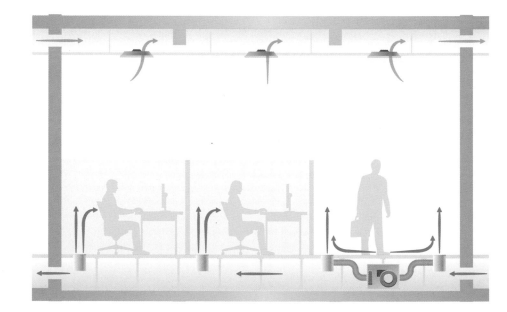

UNDERFLOOR AIR DISTRIBUTION

Underfloor air-distribution systems rely on the construction of a raised floor roughly a foot above the floor slab. A series of plinths support removable floor tiles, under which the air diffusers are typically located. Air pumped through these diffusers will generally be slightly warmer than that coming from traditional ceiling systems and can easily be controlled by nearby office workers.

Dissecting an air-handling unit

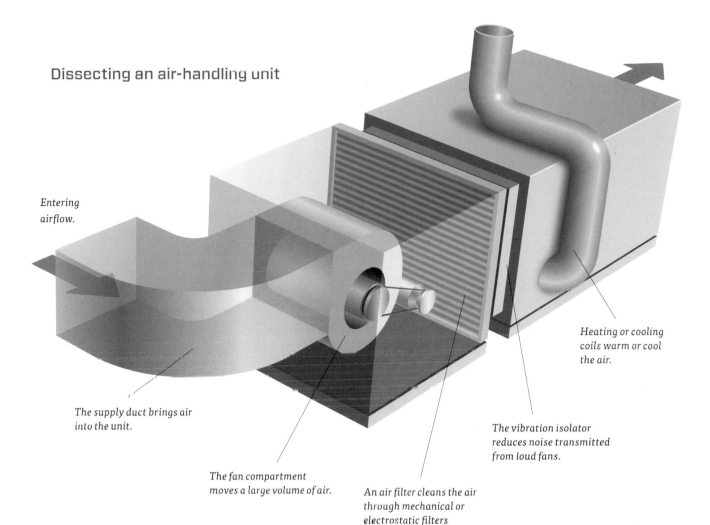

Entering airflow.

The supply duct brings air into the unit.

The fan compartment moves a large volume of air.

An air filter cleans the air through mechanical or electrostatic filters

The vibration isolator reduces noise transmitted from loud fans.

Heating or cooling coils warm or cool the air.

Wind and the stack effect

Like it or not, even the most tightly enclosed skyscraper must deal with what is known as "the stack effect." As warm building air rises, the difference in density between outside and inside air (a function of temperature and moisture differences) creates a low-pressure area that pulls air into the building through lobby doors or parking ramps. It then rises through elevator shafts or stairwells to find its way out through the top of the "stack"—including the elevator hoistway or the mechanical penthouse. The greater the temperature differential between outside and inside, and the taller the building, the more powerful the chimney, or stack, effect can be.

Sometimes, the stack effect works in reverse. Because of the high temperatures in the Middle East, for example, very hot air may be drawn in at higher floors and pulled downward through the building's core to find its way out at ground level.

STACK EFFECT
POSITIVE PRESSURE

NEGATIVE PRESSURE

REVERSE STACK EFFECT
NEGATIVE PRESSURE

POSITIVE PRESSURE

Heating

The invention of central heating did much to help popularize skyscrapers. Central boiler plants, usually consisting of a large cast-iron or steel-tubed boiler, removed the need for traditional room-by-room heating. Today smaller skyscrapers still rely on the concept of a central plant. Often these boilers will provide both heat and hot water to the building.

Larger skyscrapers rely on a less centralized approach and multiple mechanical floors. Air systems located on one of these floors generally provide heat to 10 to 20 floors at a time. They rely on a variety of fuels—typically gas or electricity.

Different systems may be used to provide heat throughout a skyscraper. In forced hot-air systems, air moves over a heated coil and is then circulated throughout a building. Hot water systems rely on water circulating through radiators that provide heat to individual rooms. Steam heat operates in much the same way, only it is steam—rather than water—that moves through building pipes and radiators.

In certain locations around the world, skyscrapers and other large buildings are able to tap into a district hot water or steam network, and therefore do not need to maintain boilers on-site. Buildings at Rockefeller Center, in New York City, for example, tie directly into the central steam system, as does the Empire State Building, the Metropolitan Museum, and the United Nations. Paris and Moscow, among other large cities, feature similar central steam systems.

Making steam

Very few new buildings today rely on steam for heating; most use low-temperature hot water systems, which are safer and easier to maintain but require pumps for moving the water vertically. However, New York City still features many steam-heated buildings.

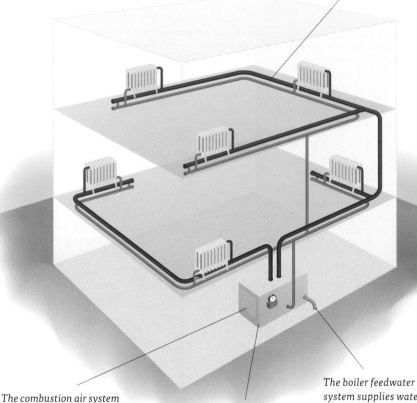

Steam piping transports steam from the boiler to the end-use services.

The combustion air system supplies the oxygen necessary for the combustion reaction and the exhaust system expels flue gases.

Fossil fuels such as oil and gas are most commonly used in the boiler.

The boiler feedwater system supplies water to the boiler. Feedwater is recirculated through the system and new, or "makeup," water is added as necessary.

District heating and cooling

In many cities throughout the world heating and cooling of large buildings is done "remotely"—i.e., the chilled water or steam is provided to the building from a central off-site source. Known as "district heating" or "district cooling" systems, they serve both commercial and residential establishments, as well as large institutional clients. Because they are only economic when multiple users can be served from a central plant, district systems tend to be found in dense urban cores or large college or government campuses.

District heating systems are popular in the colder countries of northern Europe, such as Finland, Denmark, and Poland. In Iceland, a full 95 percent of the population is served by a district heating system supported by geothermal energy. Any number of fuels can be used to power these systems, including less traditional sources such as steam, nuclear, and geothermal energy. In the United States, most district heating systems are powered by steam. New York's system is among the largest, serving 2,000 customers in Manhattan below 96th Street.

While district cooling systems are found throughout the world, they are increasingly prevalent in the rapidly developing Middle Eastern region. Midday cooling demands account for as much as 70 percent of the peak electric demand. District cooling systems help reduce this peak electric demand by shifting the load from individual building systems to a more efficient central plant. District cooling is also effective in reducing greenhouse gases: levels of CO_2 emissions and refrigerant leakage from one central plant are much lower than from many scattered plants or units.

Keeping air cool

Chillers remove heat from water via a refrigeration cycle.

━ CHILLED WATER LOOP   ━ CONDENSER WATER LOOP

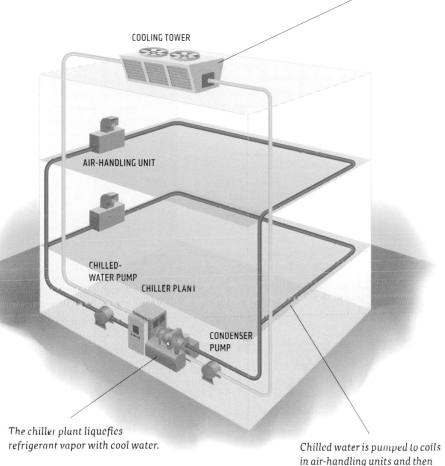

An evaporative cooling tower cools the condenser water that is heated in the chiller plant.

COOLING TOWER

AIR-HANDLING UNIT

CHILLED-
WATER PUMP

CHILLER PLANT

CONDENSER
PUMP

The chiller plant liquefies refrigerant vapor with cool water.

Chilled water is pumped to coils in air-handling units and then recirculated back to the chiller to be cooled again.

Air-Conditioning

Air-conditioning is among the largest consumers of electricity in a skyscraper. Several methods of air-conditioning are common, including a process known as "vapor absorption chilling" that—somewhat counterintuitively—uses a heat source, such as steam or hot water, to cool air. In modern skyscrapers, where a central plant is designed to serve multiple zones, the most popular technique is often water-cooled refrigeration, which relies on chilled water to cool and dehumidify air.

In a typical water-cooled system, chilled water moves through air-handling units, where it passes through coils. There air blowing over the coils is cooled before being distributed through the building. In a separate loop, cooling towers atop the building work to lower the temperature of the condenser water that is used to liquefy refrigerant vapor in the chiller.

The temperature of the chilled water moving through the system generally ranges from 35 to 45 degrees Fahrenheit (2 to 7 degrees centigrade). The temperature of the air moving across it is generally at least 45 to 55 degrees Fahrenheit (7 to 13 degrees centigrade), or at least 10 degrees higher; the differential is needed to make the transfer of heat from air to water (and transfer of "cool" from water to air) occur.

SAN FRANCISCO

In San Francisco, for example, two steam plants located downtown serve 170 customers within a two-square-mile radius in the central business district. These customers use the steam for space heating, air-conditioning, hot water, and a variety of industrial processes.

■ ENERGY PLANTS ■ DISTRIBUTION

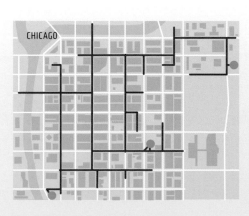

CHICAGO

Perhaps the best example of a district cooling system in the United States is found in Chicago, where Thermal Chicago Corporation owns and operates the world's largest interconnected district cooling system. Its five plants serve 100 buildings in Chicago's downtown core.

■ ENERGY PLANTS ■ DISTRIBUTION

Water

Water is critical to life in the sky. Not only is it important for drinking and sanitation, but heating and cooling in most residential and commercial office towers would be impossible without it. Today in commercial buildings roughly 60 percent of the water is used in restrooms or kitchens, while the other 40 percent is used for heating and cooling.

Historically, only one system of water delivery existed to maintain adequate water pressure for a tall building: elevated water storage tanks. These were supported by a fill pump at the bottom of the building that, when triggered, would refill the rooftop tank. Exposed to the elements, they often required some form

of heating to keep the water from freezing in particularly cold climates.

By the mid-1900s most new skyscrapers were designed with water tanks inside—rather than outside—the building. Water was pumped from the municipal system to the tank or tanks, and from there an air compressor would supply individual floors. It operated at a constant speed to maintain pressure, using considerable amounts of energy in the process.

Chicago's John Hancock Tower, designed in the late 1960s, set a new standard for water distribution in supertall buildings by featuring separate pressure zones to better meet the high flow demands of such a large building.

These zones, now common in larger skyscrapers, will either have a dedicated pump system for each zone or rely on valves to adjust water pressure for floors closest to the pump.

Today water can be brought to floors in a skyscraper through sophisticated booster systems. These systems employ "variable speed control," which automatically adjusts the speed of the water pump to maintain a constant discharge pressure. Older constant-speed systems in widespread use today maintain the same pump speed regardless of demand and rely on pressure-reducing valves to relieve pressure on certain floors.

Water wizardry

Perhaps no piece of tall building infrastructure elicits quite so much affection as New York City's water tanks. Long after the evolution of pumping technology could have made them obsolete, over 10,000 of these conically shaped tanks still grace the tops of both commercial and residential buildings in Manhattan and the outer boroughs.

The tanks date back to the middle of the nineteenth century. Because New York's gravity-fed water distribution system had only enough pressure to push water six stories up in the air, any building taller than that was required to place a water tank on its roof. Tank technology has changed little since then; when the tank water level falls below a certain point, a switch still triggers a pump at the base of the building and the tank is refilled.

Unlike most municipal water tanks, which are made of reinforced steel or concrete, New York's water tanks are made of wood—a much better insulator than either steel or concrete. The wood planks are woven together in a barrellike fashion and held together with a metal band but no adhesive at all; they become watertight as the wood swells.

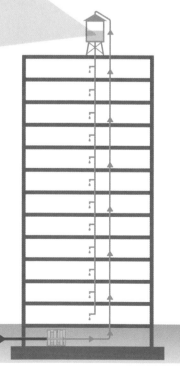

Under pressure

For extremely tall buildings, water distribution is divided into pressure zones in order to meet high flow demands. As much as 250 pounds per square inch (or psi), for example, might be required to get water to the top of the building—pressure too great to be transmitted to the fixtures on lower floors. Pressure zones are created by using pressure-reducing valves or having dedicated pump systems for each zone.

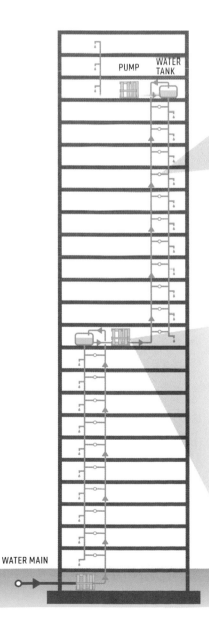

Spring

Low-pressure outlet

High-pressure inlet

Valve disk and seat

Pressure-reducing valves are compact, inexpensive devices that automatically reduce the high incoming water pressure from the city mains, booster pumps, or water tanks to a more functional pressure for distribution in the home or office. These valves also regulate water pressure by maintaining a set pressure around 50 psi.

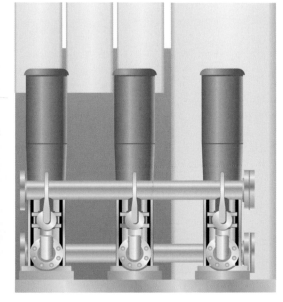

Many modern domestic water pressure booster pump systems use variable-speed control, which relies on a transducer to sense pressure and automatically adjust the speed of the pump to maintain a constant discharge pressure. This can cut energy bills in half over the system's life and significantly increase its longevity. Energy can also be saved when using constant-speed systems by incorporating a low-flow shutdown tank that receives water volume from the pump system and stores it for later use.

Drainage and Sewage

Draining water from the high floors of tall buildings should be much easier than getting it there, thanks to gravity. But it is not a particularly simple process and, indeed, gets more complex as the height of a structure grows.

As water flows down the vertical pipes designed to remove wastewater from higher floors, it adheres to the walls of the pipe—a condition known as "water sleeve." This creates a vacuum effect, pulling all other water above it in the system downward. To prevent this from occurring, air is fed into the system through a series of dedicated vent pipes, which discharge to the atmosphere to help equalize the pressure within the drainage system.

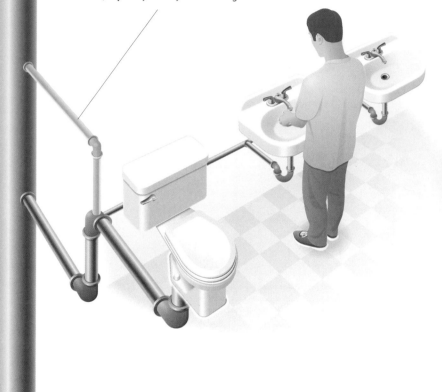

To offset pressure, relief or yoke vents slow the water before it encounters a horizontal flow change. These are located every ten floors, or even more frequently on very tall buildings.

HYDRAULIC JUMP

When water reaches the bottom of a drain stack, the horizontal drain slope may not be adequate to maintain the velocity of the water flowing vertically. Water builds up rapidly and fills the cross-sectional area of the pipe. When this "hydraulic jump" occurs and adequate venting is not in place, tremendous pressure builds up behind the jump.

The waste caravan

In some cases tall buildings have developed faster than municipal sewage systems. In no place has this been more true than Dubai, which until recently has had only one sewage treatment plant, and it has a very limited network of sewer pipes accessing it.

As a result, much of Dubai's sewage arrives at the treatment plant in a truck—delivered by tank car drivers who are paid to empty septic tanks and dispose of the contents at the plant. But there are more waste trucks than there is room to handle their cargo, and drivers can wait up to 24 hours in a queue to empty their truck and make another run.

Given such congestion at the plant, it is hardly surprising that some of Dubai's sewage never makes it there. Tank car drivers have been cited for dumping waste in open, undeveloped areas or into the storm water network; 55 such violations were found in just one week in 2008.

Infrastructure

The provision of power, water, and air to a skyscraper depends not only on the mechanical equipment at work in the building itself, but also on the municipal water and power systems that it must tie into. For most tall buildings, plugging into a reliable municipal infrastructure is no big deal; it requires a simple, although not always easy to come by, set of approvals. Where the impact of a new building upon local infrastructure is expected to be unduly large, the developer might be asked to foot the bill for an upgrading or expansion of the municipal power or water systems to ensure both adequate supply to the building and no deterioration of service to other customers.

But as tall buildings rise in new or fast-growing cities, there are increasing cases where local infrastructure cannot meet the needs of the building and its users. In some cases the supply of electricity is insufficient to reliably meet the needs of tenants, and building owners must provide extensive systems of back-up power. In others water shortages are a regular or seasonal occurrence, and the building's owner must procure water privately as needed. And in still others, direct pipeline sewage systems are either nonexistent or insufficient to handle the building's wastewater flow, and some other waste disposal mechanism must be identified.

Keep on truckin'

As one of the world's fastest-growing economies, India has, not surprisingly, witnessed a significant increase in the number of tall buildings going up in its major cities. Yet the most basic elements of infrastructure remain lacking in many of these places. Chronic water shortages in cities like Delhi, Mumbai, and Chennai, for example, require building owners to contract privately to ensure a continuous flow of drinking water to tenants—delivered not via pipe but by truck. In Delhi, for instance, 1,200 private water delivery companies charge for their services; in addition to the public sector, their clients include many of the largest buildings and fanciest hotels in the town.

India's water shortages also force tall-building owners to rely on sustainable water technologies that are optional—but increasingly fashionable—in the West. For example, "black-water treatment" technologies are used there more and more to recover as much water as possible. Water from toilets, showers, and sinks is filtered and reused in gray-water distribution systems to meet the building's toilet-flushing and irrigation needs.

Emergency power

Emergency power is necessary to ensure that exit signs, fire alarm systems, and fire pumps are always working. Other equipment on emergency power may include smoke isolation dampers, smoke evacuation fans, elevators, handicap doors, and electrical outlets in service areas. Although most emergency power systems rely on diesel or gas turbine generators, more use is being made of deep-cycle batteries or fuel cells.

COMMUNICATIONS

While the invention of elevators, central heating, and plumbing in the late nineteenth century made life in the sky possible, the invention of the telephone certainly made it easier. Getting messages to and from people atop the new skyscrapers would have been cumbersome when they reached a dozen or so floors, and next to impossible as they soared to 50 and 60 stories in the early years of the twentieth century.

Skyscrapers did not yet exist when Alexander Graham Bell, a professor of elocution by trade, invented the telephone in 1876. By converting sound waves into electrical signals, Bell found a way to transmit voices electronically from one place to another—realizing his goal of creating a "talking telegraph."

Within a year of his invention's debut, he had formed the Bell Telephone Company. Five years later most major cities had telephone exchanges operating under license from his company. By 1904, thousands of independent telephone businesses had sprung up across the country—though most couldn't talk to one another, and it was only a matter of time until they would be integrated.

While the scientific principle of using electricity signals shaped like sound waves to transmit the voice was universal, the actual equipment used to talk on the phone varied from place to place. In rural areas most telephones were powered by a "magneto" or "crank" and incorporated a battery. Turning the crank sent a signal through the wire to alert a central operator that a call needed to be placed. In contrast, systems in larger cities were powered centrally by what was known as a "talk battery" at a central telephone exchange; local users needed only to pick up the phone to request a connection.

Typically operators at a switchboard took the name of the party being called and

Data wave

In 1965, Intel cofounder Gordon Moore predicted that the number of transistors on a chip would double about every two years. Processing power has steadily risen since then because of increased transistor counts and the need for faster communication to share massive amounts of data.

MORSE CODE

1820   1840   1860

Switchboards were use to make and route calls to their destinations.

Most switchboards were staffed by women, because they were thought to be more polite.

plugged the call into a line to carry it either to another part of the office or to another office, where a second operator would contact the party. Originally these operators were young boys, many of them in their teens. But in the 1890s, these boys were gradually replaced by "telephone girls" or "hello girls"—women well versed in manners and enunciation and better able to follow a standard operator script.

For the first two decades of the twentieth century, most calls were local in nature. The first transatlantic connection was made in 1915, and for some time thereafter long-distance users had to make appointments to use special soundproof telephone booths for their long-distance calls. These calls typically took longer to place than local ones—particularly during the Second World War, when lines were

reserved for military purposes and callers could wait up to two hours for a connection.

During the same period automation came to telecommunications—and by the early decades of the 1900s nearly one million callers were dialing their own calls. Analog signals traveled through what became known as the "public switched telephone network" (PSTN), which remains the backbone of the U.S. telecom network— although signals today travel digitally from one point on the network to another.

Technological change in the field of telecommunications during the second half of the twentieth century was dramatic— leading to an exponential rather than a linear growth in usage. The conversion of signals from analog to digital for travel over distance greatly improved the efficiency of telephone systems around the world. Since then, advances in fiber optics, optical switching, and electromagnetic technology have led to greater capacity and further improvements in performance.

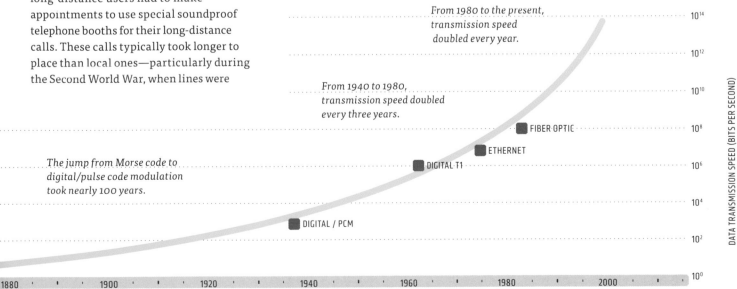

From 1980 to the present, transmission speed doubled every year.

From 1940 to 1980, transmission speed doubled every three years.

The jump from Morse code to digital/pulse code modulation took nearly 100 years.

FIBER OPTIC

ETHERNET

DIGITAL T1

DIGITAL / PCM

DATA TRANSMISSION SPEED (BITS PER SECOND)

10^{14}
10^{12}
10^{10}
10^{8}
10^{6}
10^{4}
10^{2}
10^{0}

1880   1900   1920   1940   1960   1980   2000

How Information Travels

Innovations in telecommunications must work "backward": new products must be compatible with existing telecom networks or they are unlikely to find an audience (e.g., telephones manufactured 50 years ago still work, as do first-generation home computers). For that reason most telecommunications systems or networks are hybrid in nature, incorporating a variety of both traditional and new technologies, such as cable (fiber or copper), radio, infrared, or wireless.

Traditionally telephone lines were made of copper cable: it was a better conductor than gold or aluminum and was cheap and easy to manufacture. Twisting two cable wires together into a "twisted pair" was found to create a sort of shield that minimized interference with voice conversations, and this became the standard format for a single phone line coming into and out of residential and commercial buildings.

In the 1970s, fiber optic technology—the transmission of information as light impulses along a thin glass strand—was introduced. Fiber optic cable uses glass as its primary conductor and can carry significantly more information than conventional copper wire; it is also less subject to outside interference. Because the amount of signal degradation over distance is far less than that of cable, it quickly became the norm for long-distance telephone lines around the world.

A single fiber optic line consists of hundreds of glass strands bundled within individual tubes that are together protected by an outer sheath. In outside environments these tubes are typically bundled loosely and float in a gellike substance to minimize damage from any external pulls on the cable. However, fiber optic lines within buildings are more protected and their tubes will generally be bundled tightly—i.e., the fiber strands form part of the cable structure. Within the building they will travel either vertically, in risers or shafts between floors, or horizontally, between offices on the same

PBX

The phrase "private branch exchange," or "PBX," was first applied when switchboard operators ran company switchboards by hand. PBXs make connections among the internal telephones of a private organization—usually a business—and also connect them to the public switched telephone network (PSTN) via lines from the telephone provider. Because they incorporate telephones, fax machines, modems, and more, the general term "extension" is used to refer to any end point on the branch. Dialing a number "to get an outside line" means using a PBX.

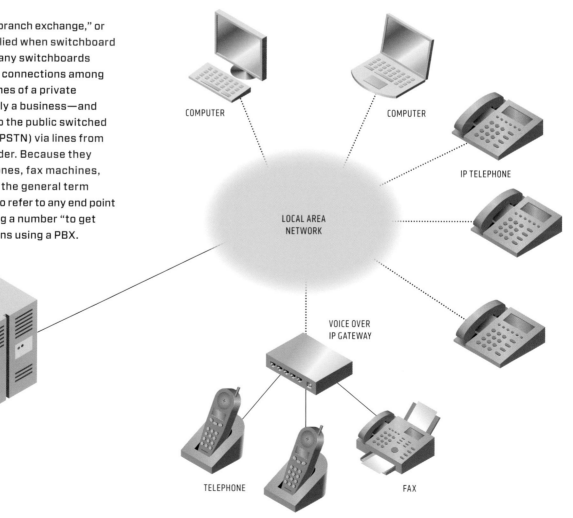

COMPUTER

COMPUTER

IP TELEPHONE

LOCAL AREA NETWORK

VOICE OVER IP GATEWAY

PBX

TELEPHONE

FAX

floor, or through the ceiling plenum between floors.

Whether copper or fiber optic, cables for communication with the outside world to offices have often been connected to private branch exchanges (PBXs)—a system on the customer's premises that manages all data and voice networks moving within a business. By handling intraorganizational calling privately, it reduces the charges for local phone service significantly. Today PBXs are increasingly taking the form of "hosted PBXs," i.e., the various internal services are provided by the local telephone company at its own exchange under lease so that the customer does not need to provide all the switching equipment.

Telecom pathways

A variety of wires run through a skyscraper to ensure round-the-clock communication. Increasingly, landlords must provide convenient access to multiple telecom providers to meet tenant demands. To ensure reliable service, some office towers will be designed to feature not one but two points of connectivity with the outside world.

- SINGLE-MODE FIBER DISTRIBUTION
- MULTIMODE FIBER DISTRIBUTION
- VOICE-GRADE & HIGH-SPEED COPPER DISTRIBUTION
- CATEGORY 5 DISTRIBUTION
- COMMUNICATION ROOM
- REDUNDANT ENTRANCES

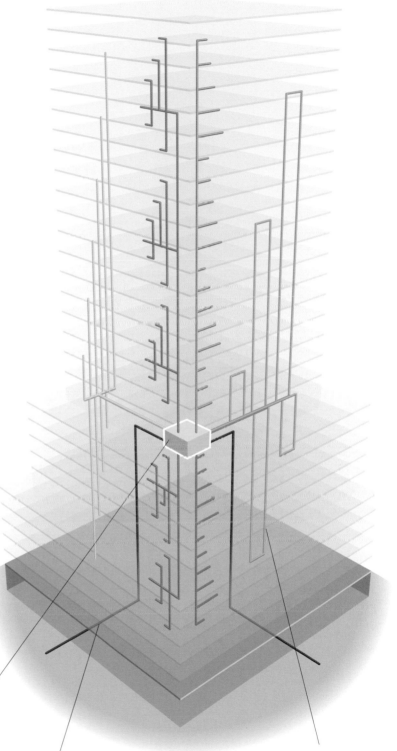

The communication room is often located in the basement, but in certain new buildings it is located on a higher floor to provide multiple paths of access to different floors.

Incoming telecom services connect the service provider and end users in the building.

Fibers may be wired in a "butterfly" form in order to provide redundancy if a fault occurs somewhere along a line within the building.

Hiding antennas

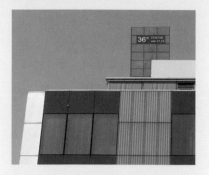

Cell towers are harder to site in urban areas than they are in rural ones. Land for a stand-alone tower is rarely available, and zoning regulations often prohibit its construction. As a result, the roofs of tall buildings in cities are increasingly becoming home to cell transmission or booster equipment.

These fixtures, which generate a small revenue stream for building owners, take various forms. They may be in an outdoor cabinet, mounted on a side wall, or located just below the roof. Often an attempt will be made to blend into the building's architecture or, even better, hide the fixture entirely from public view.

Keeping a signal alive

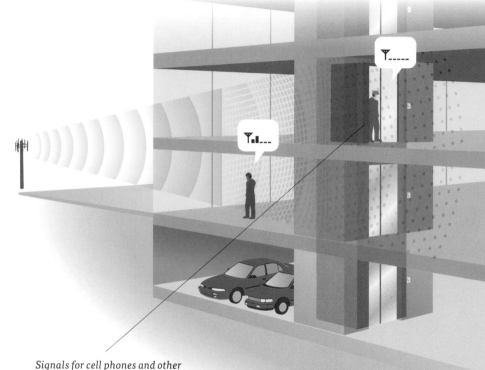

Signals for cell phones and other wireless devices are seriously degraded by passing through dense construction materials. As most people have experienced, cell reception in underground parking garages and elevators is often poor or nonexistent.

TYPICAL SIGNAL LOSS

Wireless

Over the last 20 years, wireless technologies have dramatically reshaped telecommunications. While these wireless systems have been much easier to implement in many places than cable-based connections, due to their lower capital and infrastructure requirements, they have proven somewhat more troublesome for tall-building owners as the cell signals find it hard, if not impossible, to penetrate the steel and concrete structures that support modern skyscrapers.

Cell coverage is a function of many things, including the strength and frequency of the signal, the size and power of the transmitter, the direction the signal is coming from, the intervening geography, and the weather. Even the largest cell towers have limited capacity, and the spacing of masts is generally a function of the density of the population they serve: in urban locations cell towers must be as close as one quarter to one half mile apart, while in the suburbs they can be comfortably spaced a mile or more apart.

Like all radio signals following a predetermined path, cell signals deteriorate over distance. Even when they are picked up and redistributed by cables, fiber optics, and antennas, signals will lose strength as they move farther from their source. And while radio signals do not necessarily need a clear line of sight to be effective, interposing obstacles serve to degrade or attenuate the strength of the cell signal, creating "dead zones" where service is inaccessible.

Dead zones within skyscrapers can be anywhere. In skyscrapers in urban settings, it is the lower floors that can often suffer the poorest cell coverage due to interference from nearby buildings. But skyscraper structures themselves often interfere even with direct signals from outside towers. Steel beams, brick walls, and wallboard supports can block the strongest of signals—as firefighters at the World Trade Center on 9/11 sadly found out.

To overcome this problem, many modern skyscrapers feature systems of "boosters," or "cellular repeaters," that pick up cell signals coming from outside and repeat them inside the building (and vice versa). Repeaters generally include a base station, fiber optic radio heads that feed a fiber optic distribution system, and a special antenna. They can be located as densely as needed throughout a tall building—often on every floor. Most important, they must support the wireless systems of various carriers, each of whom uses different frequencies and technologies.

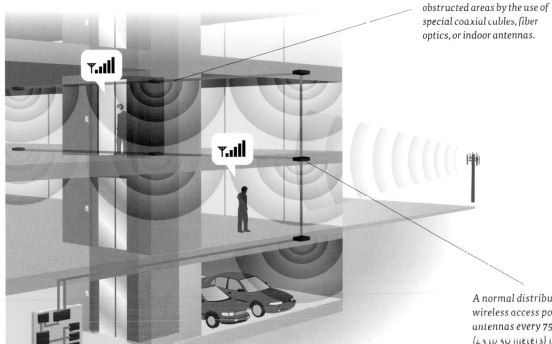

Signals may be boosted within obstructed areas by the use of special coaxial cables, fiber optics, or indoor antennas.

A normal distribution of wireless access points requires antennas every 75 to 100 feet (23 to 30 meters) to ensure adequate reception.

SIGNAL ENHANCEMENT

Some building owners use these internal networks for wireless communications where wireless handsets, or walkie-talkies, would be impractical.

Beyond radio

Most heavy construction relies on two-way radio communication among workers—or, as it is known in the United States, "walkie-talkie" communication. But as reliable as walkie-talkies have proven over time, they are no match for the dizzying heights to which modern skyscrapers aspire.

During construction of the Burj Khalifa, walkie-talkies being used by Samsung employees stopped working once construction reached the thirtieth floor. To ensure continued communication between the crane operator and the men on the top floor, a new technology that converted radio signals from mobile devices to an IP signal using VoIP (voice over internet protocol) technology was introduced. It allowed communications to travel over a wireless mesh similar to those used in mobile phone systems, but more localized.

This mesh also offered the ability to transport video. As a result, in addition to restoring full communications among workers, the new system allowed construction managers on the ground to monitor in real time the safety of workers high up on the construction site.

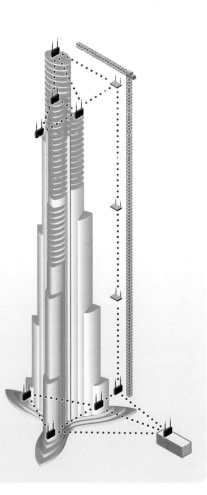

Public Safety

Telecommunications is not only important to residents and workers in tall buildings; it is also critical to firefighters attempting to fight fires. Like workers on construction sites, firefighters within buildings have historically relied on their own form of walkie-talkies—both to communicate with one another on incident floors and with battalion chiefs organizing the deployment of personnel from remote locations.

But skyscraper walls and elevator shafts have proven as troublesome for firefighter radios as they have for cell phones. In some cases fire departments have found radio bands that successfully penetrate steel-and-concrete walls. In others they have devised special devices that merge a variety of radio or cell frequencies to allow multiple city agencies, including police

and fire, to talk to one another. Some municipal fire departments have devised powerful mobile repeaters that can be carried by firefighters to strengthen the radio signals emanating from handheld devices on the fire floor.

Certain skyscrapers have more sophisticated systems for firefighters to use. Large buildings may have their own dedicated channel that provides a wireless spectrum in the event of a disaster. Others are dotted with repeaters, which can effectively bolster radio communication between firefighters. In both cases, however, these innovations are of little use if fire personnel don't know how to use them: both buildings at the World Trade Center in New York had extensive repeater systems that were not used properly on 9/11.

Command post radio

One of the most effective weapons in fighting interference from concrete and steel within a skyscraper is also one of the simplest. Shortly after reports on the 2001 terrorist attack at the World Trade Center highlighted poor communication among fire department personnel, one New York City fire captain turned his energies to inventing a better mousetrap.

He took the battery from his daughter's Jet Ski and hooked it up to an old marine radio, creating a high-powered mobile repeater that works effectively at the highest zone of the radio spectrum. A briefcase-sized device weighing 22 pounds (10 kilograms), his "command post radio" can be carried up to the fire floor to communicate effectively with fire personnel in other parts of the building. So strong is the signal emanating from the device that it comes with a nine-foot cord to allow firemen to stand away from the frequency radiation it emits.

A portable repeater can be carried to the floor below the fire to facilitate communications between firefighters on high floors and chiefs in the building or outside.

Repeaters mounted on neighboring buildings or in mobile battalion communication cars can be used to boost signals.

Disaster-speak

There is not much that people have agreed on about 9/11, from the identity of its masterminds to the nature of its memorial. But one fact is generally accepted: poor radio communication among New York's firefighters led to a greater loss of life at the World Trade Center than was necessary.

Two particular communication problems were largely to blame. The first was the degradation or obstruction of radio signals caused by the building's structural steel beams. The second was the sheer volume of radio traffic, which overloaded the system and led to communication failures; at one point, 90 radios were attempting to share one frequency. Together these issues were responsible for somewhere between one-third and one-half of all radio communication between fire personnel being unintelligible.

Cell phone service on 9/11 was also problematic as its volume spiked to levels well beyond what local carriers could handle. Yet, somewhat ironically, landlines in the World Trade Center continued to work for some time after the two planes hit the buildings. Police Department communications were also much more robust—possibly because their command center was set up more than a block away and had a direct line of sight with police personnel in the building.

LEAKY CABLE

One of the most promising techniques to fight interference from structural obstacles in skyscrapers is something known as "leaky cable"—a technology successfully used in mines, underground shopping malls, and subway and utility tunnels. Slits are cut into a long cable that is threaded through the elevator shaft of a building and held in place by vinyl-coated steel clamps. The cable acts as a long antenna, receiving signals from handheld radios (walkie-talkies) through its slits and carrying them to the lobby or other remote location.

Built-in repeaters boost handheld signals within the building.

Broadcast Towers

For almost 100 years, people have known that the crown of a skyscraper should be used for something—but it wasn't always clear what. The spire on the Empire State Building, for example, was originally designed as a mooring mast for dirigibles. It was only when that idea proved unworkable that experimentation with radio and television transmission from its mast began.

The first commercial television station broadcast from the Empire State Building occurred in 1941, and broadcasts from other locations followed. In the early 1950s, thanks to federal intervention in what had been exclusively an NBC (National Broadcasting Company) arrangement, the Empire State Building mast became a consolidated location for competing radio and television channels. By the mid-1960s a "master antenna" had been erected there and, until the opening of the World Trade Center in the early 1970s, it provided a base for nearly every television station (and many FM radio stations as well) in New York City.

Tall buildings serve as transmission centers for television and radio stations in many other cities as well. The antenna on the Willis (Sears) Tower in Chicago has broadcast signals for radio and television stations since the building opened in the early seventies. Today it serves as home to more than two dozen FM radio and several television stations.

Elsewhere, transmitting antennas can be found on a variety of tall buildings and structures, including the CN Tower in Toronto and the Eiffel Tower in Paris.

The Willis (Sears) Tower

FM radio and television stations broadcast from different locations atop the east and west masts of the Willis (Sears) Tower in Chicago, with varying amounts of transmission strength.

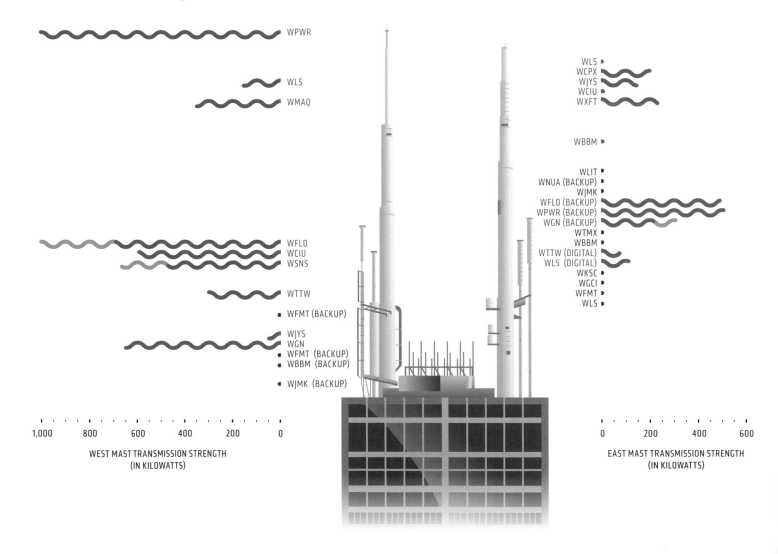

TELEVISION
FM RADIO

WPWR

WLS
WMAQ

WFLD
WCIU
WSNS

WTTW

WFMT (BACKUP)

WJYS
WGN
WFMT (BACKUP)
WBBM (BACKUP)

WJMK (BACKUP)

WLS
WCPX
WJYS
WCIU
WXFT

WBBM

WLIT
WNUA (BACKUP)
WJMK
WFLD (BACKUP)
WPWR (BACKUP)
WGN (BACKUP)
WTMX
WBBM
WTTW (DIGITAL)
WLS (DIGITAL)
WKSC
WGCI
WFMT
WLS

1,000 800 600 400 200 0

WEST MAST TRANSMISSION STRENGTH
(IN KILOWATTS)

0 200 400 600

EAST MAST TRANSMISSION STRENGTH
(IN KILOWATTS)

CHICAGO
The AT&T building at 10 South Canal in Chicago is mostly windowless.

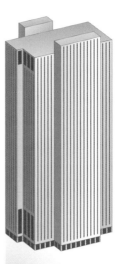

NEW YORK CITY
The 29-story AT&T Long Lines Building in lower Manhattan is largely windowless for additional security and to protect expensive telecommunications equipment.

LOS ANGELES
The 17-story AT&T Switching Center has a distinctive microwave tower on the roof, which was used from 1961 to 1993. Today the building serves 1.3 million phone lines.

KANSAS CITY
The 20-floor AT&T Long Lines Building in Kansas City was constructed in five phases over nearly 25 years and completed in 1973.

CLEVELAND
The 11-story AT&T Annex in Cleveland is a windowless switching station.

Switching Stations

Some skyscrapers devote more than just their crown to telecommunications—their entire form is designed specifically with telecommunications in mind. Known as "switching stations," they exist around the world—often with very similar heights and appearances.

Switching stations exist to house the equipment used for long-distance telephony: they make the connections and relay the signals from one place or system to another. To protect the sensitive equipment inside, switching stations generally have windowless façades, often made of concrete. Ranging anywhere from 15 to 35 floors, they are built to be stronger than skyscrapers of comparable size—able to carry anywhere from four to six times the design load of a typical office floor and to remain standing in the event of a terrorist or other incident.

Because they contain massive amounts of sensitive equipment, switching stations require proper ventilation and feature one or more mechanical floors to ensure they get it. They also must contain backup generation that can seamlessly pick up the electrical load if for any reason power to the primary switching equipment fails.

Lightning

Contrary to public opinion, lightning—the result of the buildup of a negative charge within an unstable, wet cloud—does not always strike the tallest structure around. But tall buildings are indeed more likely to experience lightning strikes than short ones, making skyscrapers—and the telecommunication and power equipment they contain—particularly vulnerable targets.

Along with broadcast masts, skyscrapers are hit more frequently than any other objects on earth. On average, a tall skyscraper will be hit between two and eight times during a given electrical storm. In Chicago and New York, with only a moderate level of storm activity, skyscrapers will experience between 50 and 110 strikes each year; in Florida, where electrical storms are much more frequent, a tall building would experience many more—up to 200 hits annually.

Although lightning can shatter windows and masonry, it is not as much a danger to humans as it is a threat to the communications systems they rely on. Direct lightning strikes can lead to the burnout of power and telecommunications equipment and to explosions of power distribution transformers. Strikes do not have to be direct to cause damage: even distant lightning can create electrical surges that move through power, telephone, or plumbing lines to a building miles away.

To protect skyscrapers, lightning rods are placed at intervals of roughly 20 feet (six meters) around the perimeter of a flat building roof or along the peaks of a sloped roof. Generally made of copper or aluminum, these strips or rods will be connected to wires running from the roof to a grounding network under the building. These wires conduct and transfer the electrical charge from the sky to the earth.

In addition to lightning rods, buildings may also feature lightning "arrestors" for additional protection. Also called "surge protectors," they are placed where wires enter a structure and provide a short circuit to the ground from the electrical conductors in a given communications or power system. By doing this they limit the rise in voltage when a particular power or telecom line is struck by lightning, thus preventing damage both to people and the electronic appliance itself.

With striking frequency...

In the United States there are an estimated 25 million lightning flashes each year, with the largest number of them occurring in the southeast and the fewest in the western portion of the country. During the past 30 years, lightning killed an average of 58 people per year.

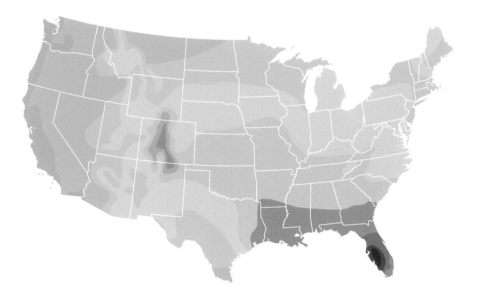

MEAN ANNUAL NUMBER OF DAYS WITH THUNDERSTORMS

■■■■■■■■■■■
10   50   100

Ground-to-cloud lightning

Lightning typically consists of a strong negative charge from a cloud reaching downward, rather explosively, to meet the positively charged ground below. One bolt of lightning will often branch off into multiple paths as it heads to earth, potentially hitting the ground in more than one place.

However, a different kind of lightning was identified in the decade after the Empire State Building opened in 1930. Called "ground-to-cloud" lightning, it originates in the

Lightning protection

A lightning flash is composed of a series of strokes (generally about four). The length and duration of each varies, but typically averages about 30 millionths of a second, with peak power per stroke averaging just over 1,000 watts.

Lightning rods capture the strike and conduct the energy to the ground.

A surge arrestor is designed to limit overvoltages on equipment. A typical surge arrestor consists of disks of zinc oxide material encased in porcelain enclosures to provide physical support and remove heat.

Low-impedance grounding rods dissipate the energy in the ground.

ground and reaches upward to the storm cloud above. It is recognizable by its distinct tree-shaped form—with a trunk on the bottom and branches on top.

Ground-to-cloud lightning is not a rare occurrence: it is responsible for the majority of skyscraper strikes. Upward discharges have been found to occur as frequently as every two to five minutes until the electrified portion of a storm region has passed, which can be up to an hour after the passage of the squall line itself.

SUPPORTING IT

LIFE SAFETY

How safe is life in the sky? The answer to that question depends where you ask it. For although the same architect may be commissioned to design office towers in New York and Hong Kong simultaneously, he or she will not design the same building for both places— and one of the biggest differences between the two will likely be in how safe they are for the occupants.

In theory the threats to skyscraper occupants around the world are similar— and not entirely different from those faced by people living or working in smaller buildings. These include extreme weather (e.g., hurricanes, typhoons, and tornados), power failures, earthquakes, and fire. Explosions, chemical releases, and terrorist attacks are much less likely to occur, but can prove very dangerous when they do.

Despite largely similar threats around the world, building codes and safety-related design requirements vary dramatically from place to place. When it comes to commercial skyscrapers, these codes are generally more conservative in certain Asian countries—

particularly in Hong Kong, China, and Japan—and least conservative in the United States. For example, taller Asian skyscrapers may be required to contain concrete cores, firefighters' lifts, floor partitions, and refuge floors. Skyscrapers in Europe share some of these requirements, while the United States has been slower to adopt many of these safety features in building codes and skyscraper design.

Most of these regulations are intended to address fire emergencies—undoubtedly the biggest threat to life in a tall, narrow structure. A skyscraper's shape is its enemy when it comes to fire: the ratio of wall to floor area and the many vertical openings, combined with the powerful stack effect on smoke and fire, mean that fire can and will spread upward quickly in tall buildings. Open-plan office floors with many combustibles, such as upholstered furniture and wooden desks, allow fire to move horizontally with ease as well. Height is also an issue as both firefighting and evacuating occupants are considerably more challenging hundreds of feet up in the sky than they are closer to the ground.

There is no evidence that fires are more likely to occur in tall buildings than in short ones; indeed, they are less likely to occur in

a commercial office or residential tower than they are in a shorter industrial building. But fires that occur high up in the sky—and the last 20 years have seen major ones in Europe, South America, the United States, and most recently China—are undoubtedly harder to control and can lead to greater loss of life.

Fortunately, most high-rise fires over the last three decades have occurred in office towers at night, when these buildings were largely unoccupied. Four floors of what was

The earliest sprinklers

once L.A.'s tallest building, then known as the "First Interstate Bank Building" (now the Aon Center), were destroyed in an off-hours fire in 1988. Eight floors of One Meridian Plaza in Philadelphia were destroyed three years later after fire broke out at night on the twenty-second floor; the damage was so extensive that the building was eventually demolished. Las Vegas has had more than its share of high-rise fires (most notably the MGM Grand and Las Vegas Hilton hotel fires in 1980 and 1981, respectively) and now has some of the strictest building regulations in the United States.

Outside of the United States there have been similarly large high-rise fires. The Parque Central Tower in Caracas, once South America's tallest building, suffered a fire in 2004 that burned for almost 24 hours, destroying one third of the 56-story building. One year later, in Madrid, the Windsor Tower fire spread from the twenty-first floor down to the third floor—destroying the entire 32-story structure. And, in 2010, a fire in a 28-story residential tower in Shanghai claimed 53 lives.

The source of these high-rise fires has varied. In some the origin was electrical—a ground fault or short circuit occurred in an unmonitored area of the building. In others the fault was human: oil-soaked rags or construction debris from internal renovation or sprinkler installation work caught fire at a time when the office building was empty.

Because fire presents so many challenges in the vertical world, most of the design and operational requirements contained in local building codes—themselves derived from national or international codes developed in conjunction with the insurance industry—are designed specifically to address it. At a minimum these codes generally include criteria for fire detection and notification, occupant egress, building materials, and fire prevention and suppression.

New York City firefighters (ca. 1908 and 1916)

Fire was a fairly common occurrence in the wooden structures of the late eighteenth and early nineteenth centuries, and highly destructive. As early as 1800 a number of efforts were under way to devise a system that would quickly douse a fire in a building—without having to wait for fire personnel to arrive with their pumper trucks and hoses.

The earliest sprinkler system was in the Theatre Royal on London's Drury Lane. It consisted of a large pipe with valves that fed into smaller pipes with half-inch holes through which water would be fed in the event of fire. This served as the forerunner to number of different "perforated

systems," which became popular in the second half of the nineteenth century in textile mills both in the UK and New England.

But none of these systems was automatic, which meant that they were of little use at night, when the mills were unoccupied and no one was available to open the water valve. The first automatic sprinkler head was invented by Henry Parmelee, of New Haven, who developed a sprinkler head containing a material that would, when heated, expand and force out a plug that held back the flow of water from a pipe system that fed it. In 1880, he installed the new automatic sprinkler system in

the Mathushek Piano Manufacturing Company's factory in New Haven.

But the real commercialization of the automatic sprinkler head was done by his manufacturer and licensee, Frederick Grinnell, who improved upon Parmelee's design and patented the automatic sprinkler head that continues to bear his name. Used first in textile mills, it eventually became a standard fixture in commercial buildings. Today the SimplexGrinnell division of Tyco continues to be one of the leading manufacturers of sprinkler heads—with very few refinements made to the basic idea patented by Grinnell in 1890.

Fire Detection and Notification

Most of us, whether we live and work in skyscrapers or in shorter buildings, are familiar with fire detection systems. While there are a variety of system types, the goal of the equipment is always the same: to alert building residents to a smoke or fire condition so that they may be prepared to evacuate the building.

In residential buildings, the most common form of fire detection is a simple smoke or fire alarm that emits a loud signal when a sensor in the mechanism is triggered. Generally speaking, residential buildings will not have buildingwide automatic systems—each homeowner is instead expected to search out the source of the fire, call the fire department, and follow standard fire safety practices.

But because commercial buildings are not nearly as compartmentalized as residential ones, and are largely empty at night, automatic fire alarm systems are required by law. These fire alarm systems incorporate a number of components, including pull or signaling boxes, fire or heat detectors, a control unit, and audible alarm devices. Many will also include visual alarm signals (often flashing red lights) for the hearing impaired as well as emergency telephones.

Alarm systems in commercial towers are generally required to have primary and secondary power sources as a fire-related emergency can disrupt the normal power supply. Secondary power, usually coming from batteries or an emergency generator on-site, is expected to come on within 30 seconds of the failure of the primary power supply.

Simple alarm systems have historically been responsible for alerting building occupants to the fire and communicating with a municipal fire department. More sophisticated systems, now in use around the world, are designed to directly control building equipment and services in the event of a fire. This can include pressurizing stairwells, releasing "hold-open" devices on fire doors, shutting down air circulation systems, activating smoke exhaust systems, and recalling elevators to the ground floor as soon as fire is detected.

SMOKE DETECTORS

Photoelectric smoke detectors use reflected light to detect smoke. Ionization detectors use a radioactive source to ionize the air between two electrodes; when smoke disrupts the flow of current, an alarm sounds.

CLEAR AIR

SMOKE PARTICLES

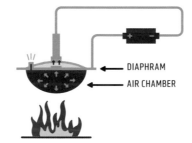

HEAT DETECTORS

Heat detectors can be programmed in one of two ways. They can be triggered either when a certain predetermined temperature is reached or when an abnormal rise in temperature occurs. Flame or "optical" detectors may sometimes be used instead: these "see" the fire by detecting electromagnetic radiation emitted by combustion.

DIAPHRAM

AIR CHAMBER

PULL BOXES

Manually activated pull boxes are constructed to be easily seen, identified, and operated. They are typically located near exits from the floor.

FIRE ALARM CONTROL PANELS

The fire alarm control panel is the "brain" of the alarm system, transferring signals sent from pull boxes or fire or smoke detectors to alarm signal devices. It can also send information to municipal fire departments as well as control fire-related equipment, such as fans, doors, and ventilation systems. Most fire alarm control units monitor themselves on a real-time basis by sending current through the wires to ensure each of the system's circuits is functioning.

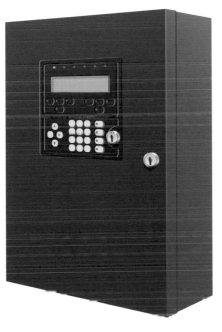

ALARMS

Alarms use audible, visible, tactile, textual, or even olfactory stimuli (odorizers) to alert building occupants. Audible or visible signals are the most common and typically use speakers to deliver instructions to people in the building.

VENTILATION

Systems can be designed to control the operation of building ventilation equipment to minimize the spread of fire and smoke. Signals from the system can engage equipment to pressurize exhaust systems in order to ventilate a fire and reduce heat buildup.

FIRE DEPARTMENT COMMUNICATION

One key output function is emergency response notification. An automatic telephone or radio signal is communicated to a constantly staffed monitoring station or 911 center, providing information about the location of the alarm.

Evacuation

Evacuation of tall buildings demands a protected and unobstructed path to the outside provided through multiple stairwells. Designers have some freedom as to where these stairwells are placed, but their configuration, number, and size are generally delineated carefully by local law. And laws differ dramatically from place to place.

Asian countries have promulgated perhaps the most far-reaching code requirements. In Japan, fireproof corridors anywhere from 20 to 30 feet (six to nine meters) long must be located on every floor as places of refuge, and vestibules leading to fire stairs must be pressurized. In China, floor partitions—which drop in the event of a fire—must be incorporated into new high-rises. Many codes in Asia also require buildings to have concrete cores and firefighter lifts—special elevators designed to transport emergency personnel to the fire floor.

Many Asian countries also require "refuge floors." Located every 15 to 25 floors of the building, often on a mechanical floor, refuge floors are designed to provide a steady stream of fresh air and maintain emergency lighting for at least two hours, giving occupants enough time to rest, transfer to another set of stairs, or be removed from the building by emergency elevators.

Not all emergencies require full building evacuation. New "defend in place" strategies—which envision most building occupants staying put or moving to refuge floors—have become part of building emergency response plans in many places.

Descending from on high

A conservative estimate of the time needed for most occupants to descend undamaged and smoke-free egress stairs is about one floor per minute (50 seconds per floor were reported in the WTC evacuation). A variety of rules govern the number, location, and nature of exit stairs.

Building codes in many places outside of the U.S. require "firefighter lifts," designed to give fire personnel easier access to high floors during fire incidents. These special elevators generally have their own power systems and are pressurized independently from other areas of the building. The lifts are not just specific to high-rises; in England, firefighter lifts must be located in all buildings over 60 feet (18 meters) in height.

An area of refuge serves as a temporary haven from the effects of a fire or other emergency when evacuation may not be desired, safe, or possible. The area is typically equipped with a steady supply of fresh outside air.

Emergency elevators

Nearly everyone knows to use stairs rather than elevators to leave a building during a fire. But given the height of modern skyscrapers, full evacuation by stairwells may be difficult, if not impossible, to accomplish in less than two hours. As a result, the fire protection community is beginning to embrace the idea of egress elevators—which can cut building evacuation times down to less than half what they are with stair-only evacuations.

The concept of elevator evacuation has led to the design of what are sometimes called "protected elevator systems." System features include water-tolerant components, a pressurized shaft and elevator lobbies on each floor, fail-safe power, special smoke-protection mechanisms, and more sophisticated ways of communicating with elevator occupants. The elevators are also designed to stop slightly higher than floor level so sprinkler water from the floor can't run into the elevator cab.

Emergency elevators have been incorporated into buildings in Asia, the Middle East, and the United States (e.g., the Stratosphere Tower in Las Vegas). Typically building occupants are expected to use stairwells to move down to refuge floors, from which they will be evacuated if necessary by fire personnel using elevators.

Where two or more exits from a floor are required, they should be separated by a minimum distance. This distance is typically a function of the size of the floor but is also partially determined by the presence or absence of sprinklers.

Every sign required for exit or exit access must be continuously illuminated and legible in both normal and emergency lighting modes. Exit signs with a directional indicator must be placed in locations where the nearest exit is not apparent.

The number of exits and the width of stairways are typically dependent upon the size of the building, the building use, and the number of occupants. At a minimum, buildings are generally required to have at least two exit routes on each story.

Each exit must discharge directly outside to a street, public walkway, open space with access to the outside, or other approved refuge area. The exit area must be large enough to accommodate the occupancy load of the building.

High-Rise Fires

Each year there are an estimated 15,500 high-rise fires in the United States alone. Most—nearly three fourths of that total—occur in residential buildings. The two most frequent causes are the exact same as they are in shorter buildings: stoves and cigarettes. But most of these are small, extinguishable fires; a full 95 percent of them are successfully confined to the room in which they originated.

The fires that are not so easily contained tend to be located in office towers, particularly those without functioning sprinkler systems. (While automatic sprinkler systems are mandated in new buildings in many places, not all municipal codes require building owners to retrofit older buildings to accommodate them.) Almost without exception, buildings with fully operational sprinklers have little to no chance of suffering a major fire in the absence of extraordinary circumstances.

The speed at which flames spread through a building depends on a number of things: the fuel it is consuming (i.e., its combustibility), the orientation of the fire, the air supply, and its surface-to-mass ratio. At top speeds fire can travel up to 20 feet (six meters) per second. In skyscrapers, the absence of compartments, or fire barriers, facilitates this travel horizontally. Likewise, the gaps between the building's structure and its external cladding allow fire to rise vertically—causing a skyscraper to act, in the absence of mechanical and physical intervention, like a big chimney.

Fire at the MGM Grand

One of the worst fires in skyscraper history happened in 1980 at the MGM Grand Hotel complex in Las Vegas. Built in 1970, the complex consisted of a 21-story hotel above a casino, a showroom, convention facilities, and restaurants. While the hotel tower itself was partially sprinklered, the other facilities—where the fire started—weren't, resulting in 85 deaths and over 600 people injured.

The fire began in a restaurant near the ground-floor casino and, in the absence of sprinklers, spread rapidly. The building's ventilation system, rather than being shut off to deprive the fire of oxygen, continued to provide fresh air to it. The vertical shafts, particularly the elevator hoistways and exit stairways, conducted the toxic smoke from the fire in the restaurant on the ground floor up to the higher floors of the hotel.

Most of the deaths were at the top of the 2,000-room hotel and were the result of smoke inhalation—not of contact with the fire itself. Many people died in their sleep or fell unconscious while trying to escape.

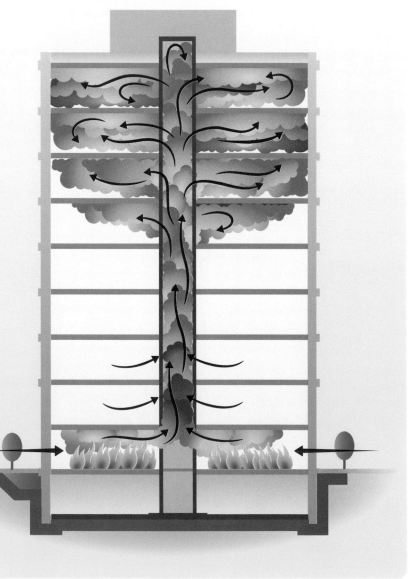

Flashover: how fire spreads

Soon after a fire has started, smoke and heat will generally collect in the upper portion of a room. The heat from the smoke layer radiates energy back to the unburned objects in the room, which get hotter and hotter. In a small space many of these objects will typically ignite at once, in a phenomenon known as "flashover." In a larger space this rarely happens, and objects generally ignite sequentially based on the material they are made of and the nature of the ventilation in the room.

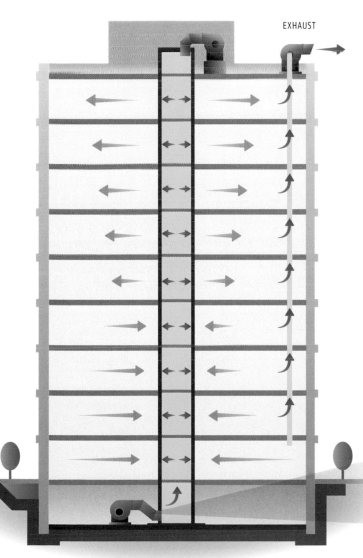

EXHAUST

PRESSURIZED STAIRWELL AND SHAFT

Stopping the spread of fire

There are four basic methods to actively or passively control the spread of smoke and heat and provide safety to building occupants, firefighters, and property:

PASSIVE COMPARTMENTALIZATION
Physical barriers (walls, floors, doors, smoke dampers, etc.) are used to hinder the movement of smoke from the fire area into the nonfire areas.

BUOYANCY
Fan-powered and passive vents, typically located in the ceiling of large, open spaces, such as atria, act to vent hot, buoyant combustion gases.

DILUTION
Fresh air is used to dilute the smoke and limit combustion in nonfire areas.

PRESSURIZATION
Fans are used to maintain the stairwell and elevator shaft at a higher pressure than the adjacent spaces to prevent infiltration of smoke into them.

Fire Resistance

When it comes to slowing down fires, materials count—that's why local building codes regulate carefully what kinds of materials can be used to construct skyscrapers. Codes will typically specify which type of material is allowed where and how long it must be able to withstand a fire (its "minimum hourly resistance rating"). For some components the goal is preventing the spread of fire, or "barrier fire resistance"; for others the stability of the surrounding structure, or "structural fire resistance," is of paramount concern. Walls, for example, typically serve both purposes.

Each component that will become part of a skyscraper is, in prototype form, subjected to full-scale fire-based tests under highly controlled conditions. In general, these tests are based on "time/temperature curves" that determine how long a product can retain its shape, as well as its strength and load-bearing capacity. However, testing protocols vary from country to country. As a result, the same product will often have to be tested in multiple countries to satisfy insurance requirements.

Following testing, components are defined based on their degree of combustibility, their performance as a fire barrier, and their structural strength. Fire resistance ratings are not solely determined by the material from which a component is made. Specified ratings can be achieved by combinations of materials, including the use of some—like cement—that are noncombustible. Assembling these components in specific, fire resistant ways can also allow them to offer greater protection.

Fire resistance is particularly important when it comes to steel, which provides the primary support for many skyscraper loads. To minimize the impact of heat on steel during a fire event, and thus extend escape times, beams and columns must be thoroughly insulated. Historically this insulation was provided by asbestos, which was sprayed on steel members following erection. Today a variety of approaches may be employed—including using other dry or sprayed products or intumescent paint, which provides a foam barrier around the steel when in contact with heat.

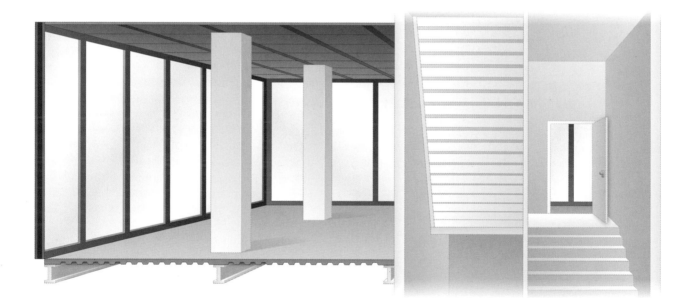

How fast will it burn?

The suitability of a particular material or component for things such as shaft enclosures, exterior walls, fire walls, and partitions, and concealed spaces is determined on the basis of fire test results. Each part of a building will have a different threshold that must be met. Exterior bearing walls, for instance, might be expected to hold their strength and shape for a minimum of three hours; interior bearing walls, columns, and beams for a minimum of two hours; and rooftops for 90 minutes.

The amount of time building components hold their strength in fire conditions is a critical determinant in the overall safety of the building. Most building codes require stairways of a sufficient number and width to permit orderly and timely evacuation of all building occupants. However, there is no agreed-upon outer limit of acceptable evacuation time, and it can easily take two or more hours to empty a tall skyscraper. The longer a building can retain its structural integrity, the lower the chance of major loss of life in a fire.

■ 60–90 MINUTES
□ 120 MINUTES
□ 180 MINUTES

Asbestos

Before it was banned in the 1980s, asbestos was the material chosen around the world to insulate and protect structural steel from the deleterious effects of fire. As a result, it is still present in older buildings in many countries. Although proven to be a carcinogen, the asbestos in these buildings is not dangerous—so long as it is not disturbed and released into the air. However, it is no longer used for fire protection and special precautions must be taken when tearing down asbestos-laden buildings.

Intumescent paint

A variety of materials are used today to perform the heat-absorbing function that asbestos once performed. One of the most effective materials to replace asbestos in the battle to keep fire from penetrating steel is intumescent paint, a surface coating that is painted or sprayed on the steel.

When a fire occurs, the coating swells up and bulges into an inch-thick layer of black foam, which becomes a protective blanket and keeps the steel from heating up. While intumescent paint doesn't stop the fire, it temporarily insulates the steel from the weakening effects of fire, and thus it provides valuable extra time for a full-scale evacuation, should one be necessary.

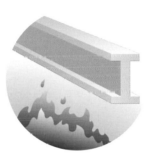

Resistance ratings

Steel, like concrete, has the advantage of being noncombustible, but its high thermal conductivity means that it absorbs heat much more quickly than other materials. The critical temperature of steel is reached between 900 and 1,100 degrees Fahrenheit (500 to 600 degrees centigrade), and only 60 percent of its original strength remains—at which point failure is imminent under design loads.

Concrete is a fairly good insulator and is usually not externally protected. However, concrete will lose compressive strength under increasing temperatures, so concrete systems must be designed with sufficient reserve strength to resist the duration of fire exposure. Designers must also ensure that the steel reinforcement is sufficiently insulated within the concrete.

STRUCTURAL STEEL
LIGHTWEIGHT CONCRETE

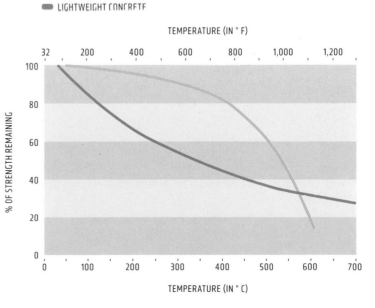

Terror Protection

While fire is perhaps the most immediate threat to inhabitants of a skyscraper, it is by no means the only one. More remote threats include plane or missile attack, airborne or water contamination, and bomb detonation.

Unlike the other dangers, bombs do not necessarily have to detonate inside the tower to inflict significant damage. Shock waves emanating from bombs on streets near buildings can create unexpected, and therefore dangerous, uplift pressure on the floor slabs of the upper stories of a building.

Most skyscrapers are not designed to resist the size of loads associated with a large-scale bomb. Nor are most smaller buildings; it would be far too expensive and impractical to construct commercial or residential buildings to deal with the unlikely possibility of a blast. However, in the wake of bombings and terrorist attacks around the world over the last decade, greater consideration is being given to incorporating features that can minimize or lessen a blast's impact both on a building and its occupants.

Some of these features fall outside of the building's footprint, along what is often referred to by security consultants as "the defensive perimeter." Unlike a military installation, which can have a significant "stand-off" or "buffer" zone, skyscrapers are often located in dense urban areas. As a result, the perimeter defenses usually consist of bollards, planters, or tasteful low walls rather than substantial fences or gates.

With respect to the building itself, there are several features now being incorporated into skyscraper design that attempt to protect occupants, who are the people most likely to be injured by flying debris. Tempered and laminated glass set

Building protection 101

For maximum safety, public parking abutting the building must be secured or eliminated, and street parking should not be permitted adjacent to the building.

Slab reinforcements can be welded directly to steep support beams so that floor slabs do not separate from steel beams in the wake of uplift forces from the blast.

One of the most effective means of protecting a structure is to keep the explosive as far away from it as possible by maximizing the "standoff distance" through use of antiram bollards or other obstacles.

To enhance protection, building columns must be designed to be sufficiently ductile to sustain both the gravity loads they are intended to support and any bending caused by the blast. When they're not, a localized failure occurs. This may cause adjacent members to be overloaded and fail, resulting in a chain of damage known as "progressive collapse." As a result, buildings may be redundantly designed to accept the loss of an exterior column for one or possibly two floors above grade without causing further collapse.

How a façade responds to any blast will significantly affect the behavior of the structure, so hardening of the façade is often the single most costly and controversial component of blast protection.

into aluminum curtain walls or steel cable nets is commonly employed as one way to absorb impact and minimize flying glass at levels closest to the blast source.

Equal emphasis is being put on design features that can protect the structure of the building itself—primarily to prevent or delay progressive collapse so that occupants have time to evacuate the building. This includes locating elements critical to the survival of the structure in the least vulnerable places, providing redundancy of critical components, and specifying structural elements (composed of both steel and concrete) capable of absorbing high amounts of energy as they bend under extreme loads.

These efforts to improve the ability of towers to withstand blasts rely heavily on science and computer modeling. Computational fluid dynamics is used to predict the loads and accompanying pressure produced by both a blast and its ensuing shock wave; computational solid mechanics is used to model the structural response of a particular building to that forecast amount of pressure.

But all the science in the world won't tell a developer how much "hardening" of a building is worthwhile. The amount of steel and concrete used to erect a building is a major factor in its cost, as well as a critical element in the design and cost of its foundation. High-strength concrete, for instance, has been used effectively to develop thinner floor slabs as a way to lighten the overall loads associated with a building; however, it is also more brittle, and therefore less likely to perform well in a blast situation. Trade-offs such as this are being made every day in the design and construction of high-rise buildings around the world.

Blast basics

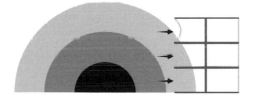

1

First, a blast wave breaks exterior windows and buckles exterior columns. Unaffected adjacent columns may be forced to carry considerably greater axial loads.

2

Then the blast wave moves into the building and forces its floor slabs upward. The weakening or failure of external columns can lead to floor collapse.

3

As the blast wave propagates farther, it may place downward pressure on podium roofs or negative pressure on the opposite face of the building.

The collapse of the twins

Perhaps no building collapse in the world was more spectacular than that which occurred on September 11, 2001, in New York City. Hit by planes fully loaded with jet fuel, the two towers of the World Trade Center erupted in fire and fell to the ground in quick succession—one within 56 minutes and the other within 102 minutes of being hit.

The planes' impacts were found to have dislodged asbestos sprayed on the beams and columns of the towers, allowing the fires to severely weaken the steel columns on the buildings' perimeters. As these columns buckled inward, the floors connecting them to the building's core became unstable and pancaked down upon one another.

STRONG ▨▨▨▨▨▨▨▨▨▨ SIGNIFICANTLY WEAKENED

WTC 2 (81ST FLOOR)

WTC 1 (96TH FLOOR)

Fire Prevention and Response

Because fire poses such a serious threat to tall buildings, skyscrapers contain an elaborate and extensive system of pipes and pumps to ensure that water can be provided quickly and easily to emergency fire personnel. The backbone of this system is a water main, or standpipe, that rises through the building and is connected to outlets on each floor. Water pumps are also an integral part of the firefighting network and ensure that sufficient water pressure is available to fight fires on high as well as low floors.

Most skyscrapers rely on what is known as a "wet pipe" system, in which pipes remain filled with water around the clock and release it the moment a sprinkler head is triggered. (In a "dry pipe" system, used primarily in places like an unheated warehouse where pipes might freeze, the pipes remain pressurized with air, and water only enters the system when the air pressure drops.) Only those sprinkler heads near the fire will be activated, which allows the system to retain maximum pressure.

Standpipe systems are usually present in any building taller than four or five floors (the height most fire department ladder trucks can reach from the ground). Typically they tie into the building's own water tanks, which provide several thousand gallons of water for everyday use but also serve as a backup in the event of a failure in the external water supply system. Because of their size, high-rise buildings need to be divided into separate water pressure zones—primarily to avoid excessive pressure on lower floors. In the United States, these separate zones are generally no more than 275 feet (84 meters), or roughly 20 stories, in height.

SPRINKLERS

Sprinkler heads release water when fire or extreme heat is detected. They contain a liquid that expands under heat, causing the glass surrounding it to break (normally at temperatures above 150 degrees Fahrenheit, or 64 degrees centigrade). Once the glass breaks, a plug in the sprinkler head is forced out by the water pressure behind it. Water begins flowing and is distributed downward evenly by a deflector plate.

FIRE EXTINGUISHERS

While fire extinguishers around the world look alike, they contain different things. Some are filled with compressed gas, which expands in the atmosphere and displaces the oxygen in the air around the fire. Others may be filled with carbon dioxide in liquid form or a dry chemical or powder (such as baking soda) that releases carbon dioxide in extreme heat. In high-rises fire extinguishers may also contain pressurized water.

STANDPIPES

Standpipes are a form of very tall fire hydrant, devoted to a specific building. The fire department plugs its pumper trucks directly into their base to funnel water to the upper reaches of a skyscraper. Because they are fixed, vertical structures, standpipes avoid the drops in pressure caused by hoses kinking. They are normally found running along or within a stairwell at the center of the building, within easy reach of all sides.

SIAMESE CONNECTIONS

Siamese connections are Y-shaped connections installed on or near the exterior wall of a building that provide inlets for fire hoses to the standpipe system inside the building. Typically water will flow from a fire hydrant through a fire engine pump to a Siamese connection, and then up into the building.

HIGH-PRESSURE PUMPER TRUCKS

Some cities have specialized equipment and personnel to fight high-rise fires. New York City, for example, has specialized high-rise units that serve the skyscraper districts of Manhattan. While conventional pumpers utilize a two-stage pump, these high-rise units have pumper trucks with a third stage that can supply an additional 500 gallons (nearly 2,000 liters) per minute.

FIRE PUMPS

Municipal water pressure is rarely high enough to get water to upper floors of a building. As a result, electric- or diesel-powered pumps are needed to increase the pressure to a skyscraper's standpipe and riser system. They automatically begin pumping when a building's water pressure drops—generally as a result of water being released in the system.

Firefighting

Unlike the special alarm and plumbing systems outlined on the preceding pages, firefighting itself is all about human coordination and communication—not about technology. Firefighting manuals around the world highlight the fact that preplanning, training, coordination, and communication on the part of emergency response personnel is key to successfully fighting a fire—wherever it is.

But fighting a fire high in a skyscraper requires even more training and coordination than in smaller structures or closer to the ground. Scaling stairways carrying heavy equipment can push firefighters' core body temperatures and heartbeats up to dangerous levels. To complicate matters, radio communications in high-rises are often problematic, due to interference from the building's structure and distance from the ground command stations. And the sheer number of firefighters required to search for occupants on multiple floors simultaneously requires unprecedented levels of coordination on the part of department commanders and chiefs.

Although specific firefighting strategies will vary from city to city around the world, certain features are common to high-rise emergency mobilization. For example, triangular communication—from an upper floor to an external command post and back to an internal command post—is a fairly common technique employed to overcome direct floor-to-lobby communications interference.

Typically most departments will establish primary and secondary search teams, often operating out of a "bridgehead," a forward fire command, usually located in a safe area one or more floors below the fire. Because crews can stay on a superheated fire floor for no longer than roughly 10 minutes at a time, a minimum of a three-team relief cycle is common. Rarely do these teams carry equipment more than a couple of floors; the building's elevators (or firefighter lifts), if clear, will be used to bring equipment to the bridgehead floor.

Fighting fire In the sky

The initial attack on a high-rise fire might consist of three fire companies—one ladder and two engine companies. Ground control will provide adequate water supply for the attack, and pumpers will aim to supply it to all standpipes. A full company will ensure lobby control of the elevators, alarm control systems, and stairway access; this sector maintains a log of all personnel going up to the fire area. Those ascending to the fire floor will be accompanied only by air tanks, hose packs, and forcible entry tools. A ladder crew is assigned to establish a ventilation sector, as positive-pressure ventilation of the attack stairwell is essential. Additional stairwells may also require positive-pressure ventilation.

A resource sector is typically established one or two floors below the fire floor. All spare air bottles, hoses, nozzles, and other equipment are staged in this area, which acts as a forward staging area for crews providing direct support to firefighting operations.

Helicopter rescue

Helicopters are occasionally used to assist in fighting fires in high-rise buildings—and not just to rescue occupants. Helicopters provide roof access to firefighters to enable them to open ventilation hatches or descend into the building. Because most high-rises can't handle an actual helicopter landing due to roofs littered with mechanical equipment, "heli-baskets" have been devised to allow multiple people to move onto or off the roof simultaneously.

As building heights grow, skyscrapers are increasingly being designed to accommodate helicopter landings. California was the first U.S. state to mandate that every new high-rise tower include a helipad on the roof. India's National Building Code proposed the same requirement, and the state of Andhra Pradesh has adopted rules mandating its implementation.

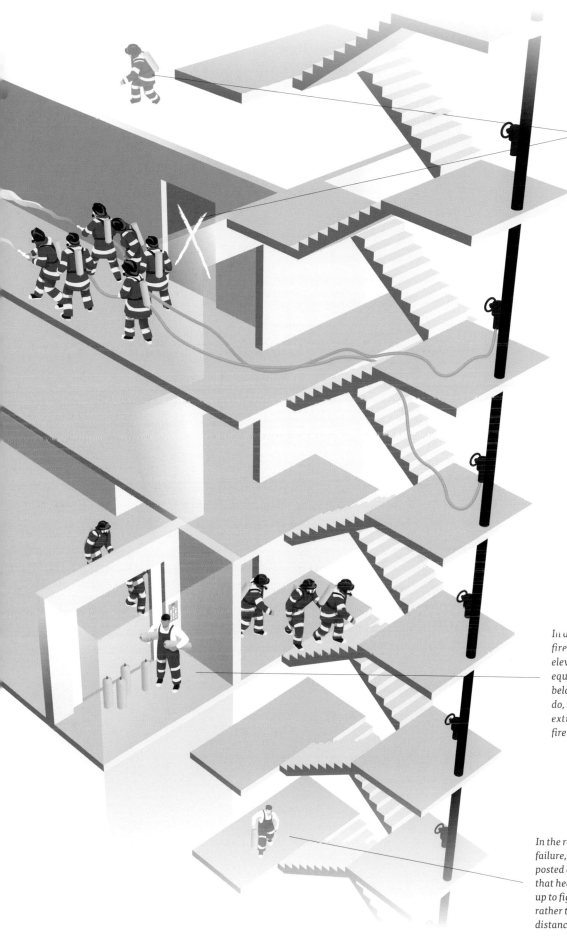

Searching teams, typically assigned to the fire floor and the floor above it, will evacuate people to locations two to four floors below the fire. A large "X" may be placed on a door to mark rooms that have been checked and emptied of human beings.

In a high-rise building, firefighters will typically use elevators to ascend and bring equipment to the staging area below a fire floor. When they do, they typically carry fire extinguishers in case they find fire at their destination.

In the rare case of elevator failure, two firefighters may be posted at every other floor so that heavy equipment moving up to fight the fire can be relayed rather than carried a long distance by a single firefighter.

MAINTENANCE

utting people in the sky not only requires tall structures to support them and mechanical systems to bring them power and water; it also requires a variety of more mundane services to look after them day to day. These services include office cleaning and trash removal, security, monitoring and maintenance of ventilation and elevator systems, and window and façade cleaning.

In residential towers, routine maintenance will generally be the responsibility of the landlord if it is a rental building, or of an organization of owners if it is not. In both cases, a third-party management firm will generally be hired to manage the building day to day. In commercial skyscrapers landlords or owners may also hire a third-party entity to operate and maintain the building or, if they own enough buildings to support a dedicated maintenance staff, do it with their own employees.

Routine cleaning is an important part of skyscraper maintenance. In residential towers the building's management will be

responsible for cleaning only the common areas. In commercial buildings the landlord will be responsible for common areas, but may also contract to clean individual office suites as well. In both cases the level and extent of cleaning is left to the owners and tenants.

In contrast, the nature and scope of other maintenance tasks is often determined by municipal regulations. Local governments usually require certain levels of façade inspection and maintenance to protect both building occupants and pedestrians on the

street below. They will also typically require regular elevator inspections and may mandate periodic building air quality testing and inspections as well.

Although the annual cost of maintenance work pales in comparison to the cost of constructing a building, over the life of a building—which can be upward of 100 years—it is much more significant. Holding aside capital replacement and reinvestment, the cost of the initial structure itself represents only about 5 to 10 percent of the total cost of owning a building over a 30-year

How old are the world's skyscrapers?

The average age of the top 25 tallest buildings in cities around the world varies dramatically by continent. Many of the skyscrapers in North America and Europe are decades into their useful life and will soon require upgrades to systems and components. The cities of Asia and the Middle East have younger building stock, much of which will not require major reinvestment for some time.

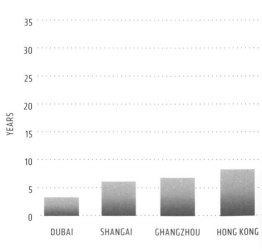

period. Operations and maintenance-related costs, however, represent 60 to 85 percent. (Land acquisition, predevelopment planning, and disposal account for the balance of its life cycle costs.)

The high operating costs of managing a building reflect the fact that most maintenance work needs to be done by people rather than machines. Experiments with automated maintenance have been tried repeatedly but have rarely been successful. The earliest automated window-washing systems, for example, were piloted in Japan in the 1980s; similar attempts in Europe followed. Most failed miserably: the initial outlay for construction of these systems was high, they were expensive to maintain, and the cleaning they accomplished was generally inferior to that done by professional window washers.

Prior to the development of glass curtain wall, window washers on tall buildings wore a harness and hooked themselves into clips on the sides of building windows. They accessed the windows from inside the floor and were thus a familiar site to office workers.

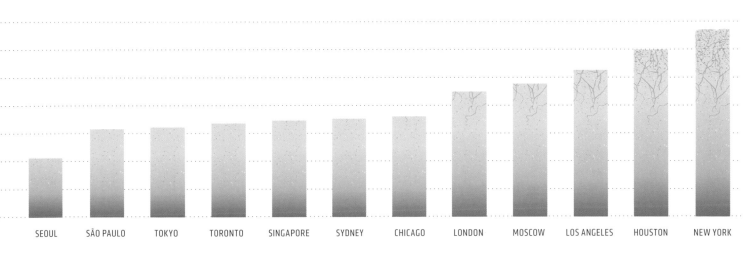

| SEOUL | SÃO PAULO | TOKYO | TORONTO | SINGAPORE | SYDNEY | CHICAGO | LONDON | MOSCOW | LOS ANGELES | HOUSTON | NEW YORK |

People and Parts

Skyscrapers must be managed around the clock. Even an office building, which is largely unoccupied after hours, must have a skeletal staff at night and on the weekends—to assist tenants who are working overtime, to oversee construction and maintenance work, or just to clean offices while employees aren't there. As a result, most buildings employ a 24-hour shift system for important services, including security, engineering, portering and front-desk coverage.

Beyond day-to-day maintenance, there is a program of annual maintenance required to ensure that a building continues to operate safely—and to provide revenue—throughout as much of its 100-year life span as possible. Periodic maintenance of a building is similar to that undertaken on a car: systems will be inspected and parts replaced regularly to protect against breakdown. But rather than mileage marking the maintenance milestones, buildings work in years: every five, 10, or even 25 years a certain part will need to be refurbished or replaced.

A day in the life of a skyscraper

Behind the scenes, maintaining a skyscraper takes a small army of people. Though the number of people employed in servicing a building around the clock will vary with building size and use, a 50-story office tower might have 20 people looking after it during the day—and many more at night.

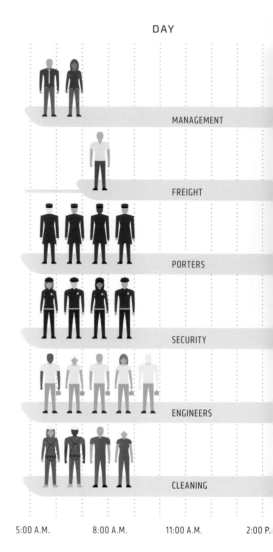

DAY

MANAGEMENT

FREIGHT

PORTERS

SECURITY

ENGINEERS

CLEANING

5:00 A.M.   8:00 A.M.   11:00 A.M.   2:00 P.

The life cycle of a building

Some parts of a building are designed to last their lifetime, such as piping, wiring, and large motors. But many other require renovation or replacement from time to time. These include water pumps and tanks, radiators, elevator components, lighting fixtures, façade and roofing systems, and ventilation equipment.

- WATER
- ELEVATOR
- HVAC
- ELECTRICITY
- FAÇADE

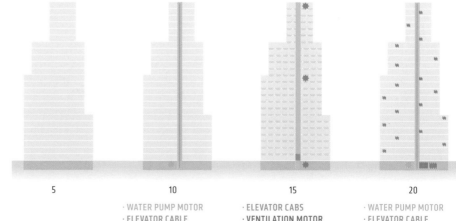

YEARS   5   10   15   20

10	15	20
· WATER PUMP MOTOR	· ELEVATOR CABS	· WATER PUMP MOTOR
· ELEVATOR CABLE	· VENTILATION MOTOR	· ELEVATOR CABLE
	· LIGHTING FIXTURES	· RADIATOR
		· CHILLER

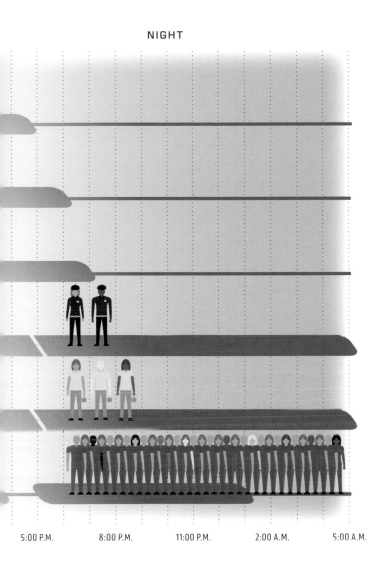

NIGHT

5:00 P.M.   8:00 P.M.   11:00 P.M.   2:00 A.M.   5:00 A.M.

MANAGEMENT

A property manager is responsible for the day-to-day operation of the building. A tenant service coordinator interfaces daily with tenants and building management.

FREIGHT

Several large-capacity elevators are operated by union-trained elevator operators and typically run from 7 a.m. to 5 p.m. unless scheduled for after-hours operation.

PORTERS

The building's common areas are serviced during business hours with a day staff of four porters; additional staffing may be provided directly to tenants for specific services.

SECURITY

During normal business hours there are four security officers and a starter on duty. After hours, one officer is posted at the main lobby desk, and building access is by secure card only.

ENGINEERS

Half a dozen engineers are typically on duty during the day to ensure smooth running of the building's mechanical system. One or two will be on duty at night, doing preventive maintenance work.

CLEANING

Three dozen full-time cleaning professionals work from 5 p.m. to 1 a.m. Specialty cleaning services, such as window cleaning and metal or stone maintenance, occur on a periodic schedule during the day.

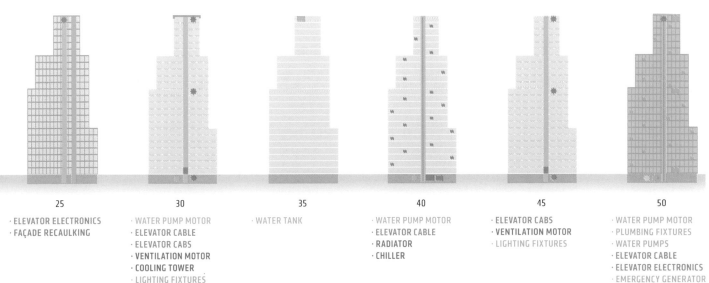

25	30	35	40	45	50
· ELEVATOR ELECTRONICS	· WATER PUMP MOTOR	· WATER TANK	· WATER PUMP MOTOR	· ELEVATOR CABS	· WATER PUMP MOTOR
· FAÇADE RECAULKING	· ELEVATOR CABLE		· ELEVATOR CABLE	· VENTILATION MOTOR	· PLUMBING FIXTURES
	· ELEVATOR CABS		· RADIATOR	· LIGHTING FIXTURES	· WATER PUMPS
	· VENTILATION MOTOR		· CHILLER		· ELEVATOR CABLE
	· COOLING TOWER				· ELEVATOR ELECTRONICS
	· LIGHTING FIXTURES				· EMERGENCY GENERATOR
	· ROOFING SYSTEMS				· CURTAIN WALL

Façade Cleaning

Skyscraper façades were arguably simpler and easier to clean when their windows opened. Indeed, façade cleaning prior to 1950 primarily meant window washing, which was carried out by a dedicated crew of workers who would strap themselves into leather harnesses and hook themselves to the sides of skyscraper windows. Though the window ledges they stood on were tiny, they felt relatively safe: if one hook failed, they would be left dangling from the other.

But with the advent of the glass curtain wall building in the 1950s, windows effectively became the building façade. And because this façade was fixed—i.e., it didn't open—window washing required access from the outside, and thus became infinitely more complex. To facilitate outside access for window cleaners, buildings had to be built with flat roofs to accommodate fixed or mobile window-washing equipment. A variety of mechanisms were developed to allow window-washing platforms to be suspended from rails or tracks atop a building roof, and to move up or down at the discretion of the operator.

Man vs. machine

Working high above the ground, safety is paramount. Cleaners must have proper training and safety equipment, such as harnesses and communications equipment.

The low-tech tools of the trade are clean water, squeegees, and dish soap that leaves windows streak-free.

Working up to 15 times faster than manual cleaning, some robots stay attached through vacuum power and do not require guiderails on the façade.

For complex façade types and areas with high wind speeds, façade-cleaning robot technology may be more suitable than manual cleaning.

Certain robots can use dry ice, demineralized water, or water with enzymes to eat away the oils and dirt. Some can also filter and recycle the used material to minimize waste.

As window-washing technology became more sophisticated, glass curtain wall buildings were able to break away from the mandatory flat-roofed silhouette. Slopes and multiple recesses, which had been prominent in designs during the first three decades of the 1900s (e.g., the Chrysler Building or the Empire State Building), began to reappear in skyscraper designs toward the end of the century. New forms of façade-access equipment were developed to serve these buildings, including arms that could act as supports for both window washing and hoisting.

Today window washing remains at the heart of a building's periodic maintenance routine. Most windows will be washed twice a year, with the exception of ground-level retail and lobby entrances, which are done much more frequently. Nearly all of this work continues to be done by hand, although new experiments with automated cleaning robots—which rely on vacuum-powered suckers rather than guiderails to attach themselves to buildings—are under way in Europe and the Middle East.

Window-cleaning mechanisms

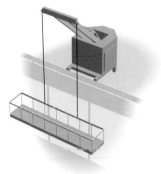

BOOM
Designed for high-rise buildings, a boom system can allow complete façade maintenance access with nothing to assemble or disassemble at the roof level.

CARRIAGES
Powered davit carriage units save the labor-intensive operation of moving portable davits. Like typical davit systems, the mast can be lowered out of view.

PORTABLE DAVITS
One of the most economical solutions for façade access, portable davit masts move between fixed davit bases and can be lowered out of sight when not in use.

BOSUN CHAIR
A powered bosun chair is designed for a single cleaner and can be operated from the chair itself.

Asian geometry

Access to the sides of a skyscraper is important not just for window washing but also for metal cleaning, inspection of curtain wall conditions, and the hoisting of replacement panels. In buildings with straight sides, this access is generally provided by a rig mechanism that is anchored on the roof and has the ability to move up or down fairly easily.

However, an increasing number of skyscrapers—particularly in Asia—do not have straight sides and distinguish themselves by twisting curtain walls or by multiple setbacks. As a result, the building designers must—at an early stage in design—select a façade access technology that will work with the building's unusual geometry.

The Petronas Towers in Malaysia offer a good example of this complexity. Because of the towers' multiple setbacks, traditional window-washing systems would be of little use. Instead, telescoping booms articulate out from hidden panels near the top of the structure, extending as far as needed to allow the cleaning rig to reach the protruding levels below.

Waste and Recycling

Skyscraper operators are responsible for getting rid of the garbage produced by people living and working within it. This includes both putrescible waste, made up of food and other organic materials, and nonputrescible waste, such as paper and cardboard, aluminum, glass, and plastic. Most landlords will rely on waste haulers—either municipal or private, depending on the city—to remove both types of waste at regular intervals. For a reasonable-sized tower, these collections will generally occur once a day, often during early morning or evening hours.

To reduce the volume of waste being transported, building operators often rely on compactors located at the base of a building—usually located near freight elevators and loading docks. By shrinking the volume of waste anywhere from 10:1 to 25:1, these fairly simple machines greatly reduce the number of trips waste haulers must make—thus lowering the overall cost of garbage removal.

Building owners want to get rid of waste as soon as they can, and not only because it takes up space. In hot climates a buildup of putrescible waste can be problematic—creating smells and attracting animals. Large office buildings may deal with this by keeping food waste in a discrete, refrigerated storage area until it is collected.

In some municipalities residential and commercial towers are required to separate recyclables. These typically include paper, glass, metal, and plastic. Recycling, like compaction, reduces the cost of traditional waste disposal by reducing the tonnage taken to landfills or incinerators.

The path(s) of American skyscraper waste

Waste from American skyscrapers typically finds its way to a variety of resting places. Much is recycled, including large volumes of paper, aluminum cans, and glass bottles. A certain amount may be burned in incinerators or waste-to-energy plants. Most of it, however, will find its way to landfills located in less urban areas outside of the city—often in neighboring states.

If trash is not picked up on a daily basis, it may be held on-site in a compactor that can reduce trash volume to as little as 1/25th of its original size. Compactors are also useful in helping to discourage scavengers, reduce insect and rodent problems, control odor, and reduce fire hazards.

Public- or private-sector trash haulers pick up trash and recycling, and sometimes both. In New York City, for example, most trash is hauled away from office buildings by private companies that pick up yesterday's trash at night; trash from residential towers is collected during the day by the city sanitation department.

A stack of paper

In recent years, almost 75 percent of high-grade office paper used in the United States has been recycled. On average, each American recycles about 340 pounds (154 kilograms) of it per year. For a very big building, regardless of whether it's residential or commercial, that's a lot of paper.

The Willis (Sears) Tower in Chicago provides a case in point. In 2008, workers there recycled 445 tons (400 metric tons) of paper. That's a stack of paper each week as tall as the tower itself—and a savings of roughly 7,000 trees over the course of a year.

Over 10 percent of the waste in the United States is directed to incinerators or waste-to-energy plants. As much as 55 percent of the waste in countries like Denmark and Sweden heads to such waste-to-energy plants.

Over 50 percent of trash in the United States is directed to landfills. Some of them, like the now-closed Fresh Kills landfill in New York, are in the middle of metropolitan areas; others require highway or rail transport to reach more remote locations.

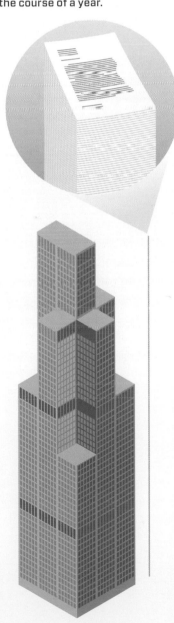

About 33 percent of all waste in the United States is recycled—much of it paper. One ton of paper from recycled pulp saves 17 trees, 7,000 gallons (26,500 liters) of water, and 390 gallons (1,500 liters) of oil—and prevents 60 pounds (27 kilograms) of air pollutants.

Cleaning

Anyone who has ever had to work late in an office tower is familiar with the arrival of the office cleaner and his or her pushcart—generally early in the evening. Yet rarely if ever does one see them leave. That's because their shift typically continues until midnight or beyond.

The cleaners who wander through office mazes at sundown are typically "generalists" responsible for undertaking a set of cleaning duties in a specific area or zone during a shift that usually lasts seven or eight hours. Other cleaners, known as "specialists," will be responsible for heavier cleaning during that same period—taking care of things like the bathrooms or the carpets.

The amount of cleaning that goes on in an individual skyscraper can be enormous. In one year alone One Penn Plaza, a large commercial skyscraper in Midtown Manhattan, went through seven million sheets of paper towel and 850,000 trash bags. Cleaning costs in that building were forecast to exceed $6 million in 2010.

Although office cleaning is still very labor intensive, it has enjoyed its share of technological improvements over the years. The invention of microfiber cloth in the 1990s permitted dramatic improvements in the quality of office cleaning. Instead of simply pushing dirt and moisture around the surface, microfiber removes the dirt and water—reducing bacteria levels by three times the amount of traditional cotton rags.

Another invention that has found a home in office cleaning is deionized water. Traditional tap water is full of minerals (such as calcium, fluoride, chloride, and silica) that bond with minerals in glass and can leave streaks. Deionized water undergoes a filtration process that removes water's negative charge; to stabilize itself the water will then seek out negative particles in the form of dirt or other contaminants and capture them—leaving windows streak-free and reducing the need for harmful chemicals.

Green cleaning

Green cleaning is a term that describes a growing trend of using cleaning methods with environment-friendly ingredients to preserve human health and environmental quality.

MICROFIBER   COTTON

Microfiber cloths are made of the same material that is used as insulation in sleeping bags and athletic clothing. They can hold up to seven times their weight in water and absorb oil and hold dirt much better than cotton cloth.

Although deionized water costs a few pennies more per gallon than normal water, it can be used without any detergent. It seeks out the dirt in ways that regular water cannot—leaving surfaces clear and free of streaking.

Vacuums with HEPA filters, which are composed of fiberglass, reduce airborne particulates by trapping them; in addition to cutting down on allergens, HEPA filters trap flame retardants and other irritants.

A large office building will use hundreds of thousands of rolls of toilet paper per year, some of it most likely previously recycled.

Ensuring air quality

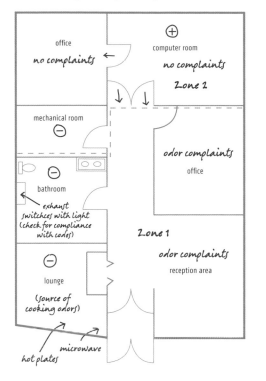

An air quality investigator will use pollutant pathway maps to understand airflow patterns and identify contaminant sources in the building.

A vacuum pump with a known airflow rate draws air through collection devices, such as a filter (which catches airborne particles), a sorbent tube (which attracts certain chemical vapors to a powder such as carbon), or an impinger (which bubbles the contaminants through a solution in a test tube).

Elevator "physicals"

Regular elevator inspections focus on the condition of the shaft and sheave, the lubrication of the governor, the doors on floors, and the various relays and switches. In New York City, every five years the governor is calibrated and the counterweight balance is checked. Tachometer readings are done to check the elevator's speed, and the governor is tripped at top speed with a full load to ensure a safe stop.

Testing and Inspection

Each year various systems within a skyscraper are inspected to make sure they are in working order. These range from the heating, ventilation, and air-conditioning systems to the elevators and standpipes. Local building regulations, often derived from national or international standards, govern what must be inspected when and set minimum standards of performance.

Elevator inspections are typically mandated by such municipal codes. In some places these inspections are carried out by contractors engaged by the local building department; in other places the building's owner is asked to hire a licensed elevator contractor to perform the requisite inspection.

The nature of periodic inspections of elevators varies somewhat from city to city, In New York City, for example, three inspections are required over a period of two years and five inspections are required over a three-year period. A no-load safety test must be done every other year, while a full-load test—including running the elevator into the buffers at the bottom of the shaft—must be carried out only once every five years.

Air quality in tall buildings can also be subject to local regulation. With so many human bodies in an enclosed space, a supply of fresh air is important to dilute the large amount of carbon dioxide being produced by people in the building. Municipal codes, generally based on national or international standards, specify how many cubic feet per minute of outside air must circulate per building occupant.

The circulation of fresh air is also important to minimize the possibility of contamination—which can occur as a result of biological growth within the building (bacteria or molds) or of chemical products

used inside or outside (e.g., adhesives, copy machines, or pesticides). In a high-rise structure such contamination—while unlikely—could potentially spread rapidly from a building's mechanical equipment through its ventilation system to multiple floors and large numbers of people, as Legionnaire's disease did some years back.

Referred to as "building-related illness," this very real form of contamination is different from what is referred to as "sick-building syndrome." Common symptoms of sick-building syndrome include respiratory complaints, irritation, and fatigue. Despite many reported cases, no causes of this syndrome have ever been found, and symptoms have generally been attributed to either environmental stress (heat, light, or noise) or to low levels of multiple pollutants in the building's air, from carpets, or other chemical-laden products.

The Elements

Regardless of whether a skyscraper finds itself in a temperate or tropical zone, it is guaranteed to be subjected to the elements in a way shorter buildings are not. The sheer size and exposure of a tall building ensures that it will be less protected from sun, wind, rain, and snow. As a result, building designers and owners must undertake a variety of protective actions to ensure that neither the building nor its inhabitants are exposed to weather-related risks.

Managing rainfall is perhaps easier than managing other forms of precipitation. Adequate roof drainage must be provided not only on the roof, but also on any building setbacks where water could accumulate. Specially designed systems are installed to ensure the controlled flow of drainage on tall buildings, as the management of flow rates and water pressure is more complex at great heights than it is with shorter drainpipes.

Snow can be more problematic. It can collect in large quantities in wind-blown drifts, placing great weight on certain parts of skyscrapers. Icicles can also be a danger during snowy weather. Each cubic foot of snow (.02 cubic meter) contains between two and three gallons (7 to 11 liters) of water. As the warm air in a building rises to melt the snow atop it, water runs down to the edge of the roof and freezes—often in the form of icicles, which hang down from ledges at great

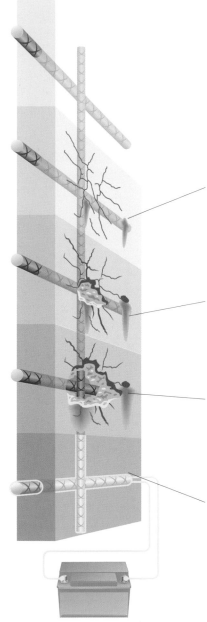

Fighting corrosion

1

Chlorides, air, acid, and water penetrate the concrete surface. It can take anywhere from 10 to 15 years before the chloride ions and acid contaminants penetrate all the way to the reinforcing steel.

2

As the rebar corrodes, rust forms and can take up five to 10 times more volume than the original steel. This process puts forces on the concrete and ultimately leads to cracking, chipping, and delamination.

3

If untreated, the process will continue until the concrete becomes structurally unsound.

4

Cathodic protection works by minimizing the difference in electrical potential between parts of the steel by introducing a constant direct current from an external source—a technique that has proven highly effective in protecting underwater steel pipelines and ships.

Falling ice

Icicles falling from buildings are by no means a common occurrence, but they can prove a fatal one. At least one passerby was killed in Chicago in 2000 by a microwave-sized icicle falling from a building downtown, and fear of its happening elsewhere has led to the development of systems to prevent icicle formation on roof ledges.

Much of the work has been done in Canada and the Midwest, where long, tough winters are the norm. One system relies on hot water pipes inside the metal framework of the building to discourage ice formation. Another involves "deicing cables," which create a heated drain path on the roof for meltwater. Still a third involves the placement of heated "ice-melt panels" around the edges of a building.

But the most popular, and arguably most effective, method of avoiding injury to pedestrians walking at the base of skyscrapers is by far the simplest. "Caution, Falling Ice" signs are regularly seen on the sidewalks of Toronto, Boston, New York, Chicago, and other vertical cities in the north—reminding passers-by that sometimes there is no outsmarting Mother Nature.

heights and can easily be blown off or fall.

Although buildings are built to last a century, their façades rarely do. Subjected to years of heat, snow, rain, and ice, curtain wall failures are common in older buildings and are usually the result of leaks or cracks that develop along surface seams over time. These cracks allow air and water to seep in, leading either to mold growth inside a curtain wall system or to deformation

of the façade framework as a result of freezing and thawing.

In some cases marble panels have deteriorated to the point where they are no longer able to support their own weight. The Aon Center in Chicago and First Canadian Place in Toronto, both designed by Edward Durell Stone, replaced their façades after marble panels weighing hundreds of pounds fell to the ground. Similar recladding was undertaken on the Gulf & Western and the Verizon buildings in New York in 2006. Both

were empty at the time, although recladding can also be done with tenants in place.

Steel-framed buildings clad in masonry can be particularly impacted by weather. Salt, moisture, and humidity can all lead to the formation of "corrosion cells" in reinforced concrete, which can lead to chipping or breakage. To protect against this, various forms of "cathodic protection," involving the introduction of an electrical current to prevent corrosion, have been developed.

A skin transplant

Recladding of a building is not always done for purely functional reasons and does not always involve replacement of the original curtain wall or skin. In some cases a glass curtain wall is erected on the outer side of a brick or masonry façade that stays in place—both to protect it from the elements and to give the building a more "modern" look. The by-product is a better-looking building—and of course a better insulated one as well.

In order to hang a new skin on an old one, brackets are drilled in the existing brick or masonry structure at regular intervals.

Glass panels are then lifted and hung seamlessly from the brackets.

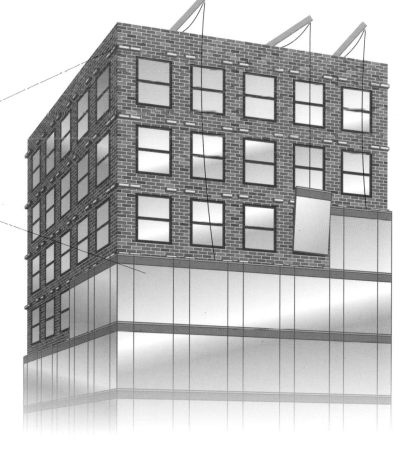

The former Newsweek Building, at Columbus Circle in New York City, shown in the photo above, has recently completed a glass face-lift of this type.

SUSTAINABILITY

The environmental movement can boast many accomplishments over the last three or four decades, but it has not changed one basic fact about modern life: people spend the vast majority of their time indoors. Americans, for example, spend roughly 90 percent of their time in an enclosed space— usually one they live or work in.

It is therefore not surprising that buildings account for roughly 40 percent of the world's energy use. Carbon dioxide emissions from buildings are significantly higher than they are from transportation; in the United States they are responsible for over a third of all greenhouse gas emissions, while transportation accounts for only one quarter. In metropolitan areas buildings hold an even larger share of the blame: they account for somewhere between 50 and 80 percent of these emissions in large cities like New York and London.

Predictably, big buildings use more energy and produce more emissions than small ones. This makes the concept of a "green" skyscraper confusing at best and an oxymoron at worst—especially as it applies to modern skyscrapers, which are largely clad in poorly insulated glass and rely heavily on mechanical, rather than natural, ventilation.

To understand just how "green" a skyscraper can or cannot be, consider the evolution of the form over the last 100 years. The earliest skyscrapers, in the late nineteenth century, had tiny carbon footprints: both air-conditioning and fluorescent lighting had yet to be invented. Their walls were thick, with a high degree of what is referred to as "thermal mass," so they stayed cool in the summer and warm in the winter. They were also bulky, with a relatively low ratio of surface area to overall building volume.

Reclaiming an energy-efficient past

| 1890 | 1900 | 1910 | 1920 | 1930 |

▮ 1885–1916

With thick masonry walls and only 20 to 30 percent of the building's surface comprised of window area, these bulky, compact buildings exhibited excellent energy performance but were heavily dependent on artificial lighting.

▮ 1916–1951

The slender buildings of this generation suffer from poor thermal performance due to their high surface-area-to-volume ratios. Like their predecessors, only 20 to 30 percent of the façade was transparent, and thus they too relied heavily on artificial lighting.

That bulkiness disappeared in the 1920s, a casualty of New York City's 1916 zoning law and architectural fashion. Slender towers gradually replaced the blocklike configuration of earlier skyscrapers; the new form allowed natural light to penetrate deeper into the floor plate. While it created more surface area as compared to volume than had previously existed, outside walls of stone, brick, or plaster remained thick and insulating, and windows still accounted for only 20 to 30 percent of building façades.

The dramatic change in skyscraper form, from an environmental perspective, came midcentury with the debut of the modern skyscraper—or what one commentator referred to as the "hermetically sealed glass box." The percentage of glass on building façades rose from 25 percent to anywhere between 50 and 75 percent, leading to a dramatic fall in thermal performance—i.e., heat losses in winter and excess solar gain in summer. Air-conditioning, of course, compensated for these swings in temperature within a building's four walls—but greatly increased its consumption of energy.

Demand for lighting increased as well. Many of the new skyscrapers were bulky in shape and featured tinted glazing, often bronze or black. As a result, they transmitted natural light poorly to the interior—leading to a higher reliance on artificial light. Their dark color also absorbed solar heat in summer, creating additional demands upon the air-conditioning and ventilation systems.

On the heels of the energy crises of the 1970s, skyscraper designers began to move away from these dark, bulky monoliths and toward new forms of clear glass with better properties of insulation. But it would be two decades before real attempts were made to design energy-conscious skyscrapers—and then it would be Europe, not the United States, that would take the lead in "green design," by focusing once again on maximizing natural light and ventilation.

Today Europe continues to set the standards for green buildings, thanks to both high energy costs and tough regulations imposed by the European Community on its 29 member nations. Among other things, the EC has imposed depth restrictions on building floor plates (assuring all workers access to natural light) and proposed strict energy-use standards for new buildings. While these and other regulations mean that European skyscrapers are unlikely ever to rival their counterparts in the Middle East or Asia in terms of overall size, they also suggest that Europe will remain the world's leader in sustainable design for some time to come.

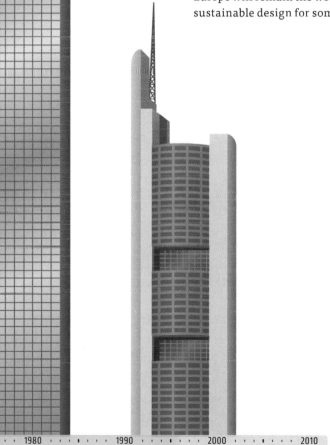

1950 1960 1970 1980 1990 2000 2010

■ **1951–1970**
The compact, glass boxes built in this era benefited from shapes with lower surface-area-to-volume ratios, but thermal performance was undermined somewhat by dark, single-pane glass façades. Fluorescent lighting and air-conditioning became prevalent.

■ **1970–1995**
Since the mid-1970s designers have looked for ways to save energy through more advanced curtain wall systems, natural ventilation, and daylighting strategies. Trade-offs between energy-efficient compact designs and opportunities for natural lighting characterized the design of slender towers.

■ **1995–PRESENT**
Regarded as the world's first "green skyscraper," the 56-floor Commerzbank Headquarters in Frankfurt opened in 1997. At nine different levels the building façade opens to sky gardens, which provide natural light and ventilation to building tenants.

Energy Performance

Buildings are massive users of energy, in aggregate consuming somewhere around 70 percent of the electricity load in the United States. On an individual building level, electricity is not surprisingly one of the largest costs building owners face; it's responsible for almost one third of their annual operating cost. So improving the energy performance of buildings matters—both from an environmental and a cost perspective.

The energy performance of a building begins with its conceptual design, which should attempt to minimize temperature extremes. This is not a new idea: in the late nineteenth century certain buildings were designed with roof ventilators or underground air-cooling chambers to regulate indoor air temperature. Likewise, in the early twentieth century buildings like the Flatiron and the New York Times buildings in New York were designed with deep-set windows to minimize solar gain.

Perhaps the earliest design issue to be considered with respect to regulating the climate's impact on a building is its siting, and in particular its orientation with respect to the movement of the sun. Known as "daylighting," this involves consideration of the configuration or shape of the building, as well as the size and positioning of windows and the height of the floors. Collectively, these factors have a significant effect on the amount of natural daylight that will penetrate the interior, and hence on the amount of energy a given skyscraper will consume over its lifetime.

Light brings with it heat, and the goal of maximizing daylight is usually tempered with an attempt to minimize thermal gain—the amount of heat transmitted into a building by sunlight.

Solar protections, such as awnings or ceramic rods on the façade, are sometimes used to reduce the sun's impact; they may be fixed or adjustable, based on the location and strength of the sun. Green roofs can also play a role, though the limited size of skyscraper roofs circumscribes their ability to reduce thermal gain.

Wind can also play a role in minimizing a building's energy demands, both by cooling the surface of the façade and by facilitating natural ventilation. The shape of a building has a direct effect on the behavior of the wind around it. The Swiss Re Tower in London, affectionately known as "the Gherkin" due to its unusual elliptical shape, generates significant air pressure differences along its façade that facilitate natural ventilation throughout the building. Other buildings have incorporated "wind scoops" into their designs for a similar effect.

Solar positioning

How a building is positioned on a site can have a significant impact on its thermal performance. Key considerations in optimizing the positioning and orientation on the site include altitude and sun path, seasonal variations in solar gain, prevailing wind, humidity, vegetation, and land contour.

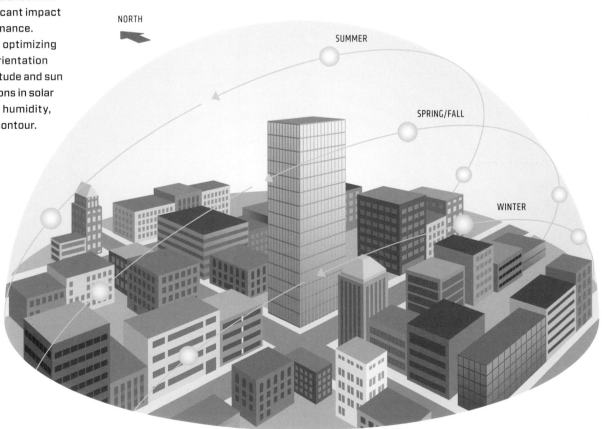

NORTH

SUMMER

SPRING/FALL

WINTER

Green roofs

A roof can interact with solar radiation to reduce both solar heat gain in the summer and heat loss in the winter. In the case of a green roof, a soil and vegetation layer shades the roofing membrane—thus reducing heat gain through the roof by almost 100 percent. In winter dormant plants add thermal mass and provide a barrier that prevents some of the warm air from escaping through the roof.

The roof's soil and vegetation layer absorbs and filters rain. The vegetation cover also adds green space to areas that otherwise would remain impervious and uninhabitable to birds, butterflies, and small wildlife.

During low-intensity rainfall (one inch or less), a green roof can largely eliminate runoff; during heavy downpours, it can reduce peak flow rates on storm water systems.

The main components of a green roof are insulation, a moisture barrier, waterproofing soil, and the plants themselves.

Urban heat island

Most climatologists are familiar with the "urban heat island" effect—the tendency for metropolitan areas to be several degrees warmer than their surrounding suburbs. This temperature difference tends to be larger at night than during the day, greater in winter than in summer, and most apparent when winds are weak.

There are two main reasons for the heat island effect. The first is the nature of man-made surfaces in cities: paved areas and dark roofs absorb far more sunlight and reemit more heat than natural vegetation. The second cause is waste heat produced as a by-product of the consumption of energy by machines, motor vehicles, and buildings.

While the effect's precise impact on the environment is unclear, scientists generally agree that it decreases air quality by increasing the production of ozone. It also has a negative impact on water quality, warming the water that flows into local rivers and thus endangering local aquatic life.

Skyscrapers are worse offenders than other buildings. They provide many surfaces for the reflection and absorption of sunlight (the "urban canyon effect"), increasing the efficiency with which urban areas are heated. The absence of natural ventilation leads to heavy reliance on air-conditioning, producing significant amounts of waste heat. Tall buildings also block more wind than shorter ones, leading to less natural cooling by convection.

Building Operations

Design is one critical factor in determining the volumes of energy that will be required by a skyscraper for heating, cooling, and lighting, but other factors are also important. Technology and systems employed within the building—often referred to as "building management systems," or "BMS"—can have a significant impact on the amount of energy a building uses over the course of a year.

Lighting is perhaps the easiest area to address as part of building management systems, and one of the most important, as it accounts for between 15 and 40 percent of a building's energy consumption. A variety of systems can now automatically control the amount of artificial light being used in a given building. Dimming systems rely on sensors to adjust levels of electric light based on the amount of natural light coming in from windows, while motion detector systems automatically turn off lights when occupants leave a room.

Improved heating, ventilation, and plumbing controls also can lead to reduced building energy usage. New air-handling controls allow more efficient delivery of desired temperatures to particular locations within a building, thereby reducing space conditioning costs. More sophisticated motors can vary both the speed of water being pumped to the higher floors of a skyscraper and the rate at which it circulates.

Finally, better technology for analyzing energy usage in all areas is now available. Advances in submetering technologies mean that tenants can access real-time information about the volume of their energy usage, often through the Internet. Combined with smart-grid technologies that promise to allow users to control building systems remotely, this could lead to significant reductions in energy usage in both old and new high-rise towers.

Automation around the clock

DAY

Building management systems (BMS) control ventilation fan systems, fresh-air mixing dampers, heating/cooling coils, and humidification systems to maintain occupied spaces at the desired temperature and humidity.

BMS maintain the indoor air quality through atmosphere-monitoring systems. As occupancy levels change, carbon dioxide–monitoring systems will send a signal to adjust fresh air intake into the building.

BMS control complex heating and cooling systems, including chilled water pumps, cooling tower controls, variable speed controls, and boilers for heating.

NIGHT

Fire alarms interface with ventilation and space-conditioning controls to shut down systems to minimize fire spread or switch them to a smoke-control or -purge mode.

BMS lighting control delivers 30 to 50 percent energy savings through programmable dimming and occupancy sensors that sense the time of day, natural lighting, and occupancy of the space.

BMS can interface with elevator controls to appropriately respond to emergency situations detected through other systems.

Substations consist of controllers, network-interface devices, and operator interfaces providing database management, communication, programming, and other user interfaces. A host computer is used to pool status data from different substations.

Demand response

Demand-response systems, which involve an electricity customer reducing its electricity use during peak demand periods in response to price signals or other incentives, are an important way to reduce energy use and costs within a building. Various technologies can be used—from smart meters and thermostats to dynamic lighting controls and energy storage systems.

In some places, large commercial users may join a peak-load reduction incentive program operated by a local utility, receiving regular payments for joining the program and episodic payments for participating in peak-demand events. Often a third-party demand-response provider acts as an intermediary between the user and the power company, developing a "curtailment plan" and notifying the client when the utility experiences a peak-demand event. Manually or automatically, the client then undertakes curtailment—which might include a lowering of thermostat set points, a reduction in lighting, or, in hotels, for example, a temporary reduction in laundry volume.

Alternative Energy

Nearly all skyscrapers rely heavily on power provided by local utility networks. Most of this energy comes from fossil fuels or nuclear plants, although in certain cities district heating or cooling based on steam, a by-product of electricity generation, is also available. With the exception of places with spotty utility performance, the only electricity that historically has been generated on skyscraper sites is the diesel-based emergency power required to keep vital building functions operating in the event of a local power failure.

But skyscraper developers are increasingly interested in meeting a portion of their energy needs at the building itself—either through cogeneration, solar, wind, geothermal, or fuel cell fixtures. The motivation for these initiatives is not, in most cases, cost; renewable technologies such as these continue to carry a higher price tag than traditional forms of energy purchased from local utilities, even when factoring in subsidies provided by government. Instead, the motivation is usually pragmatic in nature—either to comply with local regulations or to reduce the carbon footprint of a massive new building, and hence increase its attractiveness to both tenants and local authorities.

With the exception of cogeneration, few of these initiatives to produce power on-site have resulted in the benefits they forecast. Some are too new and too localized to evaluate, but in general, these attempts to reduce reliance on local power networks have resulted in neither the power savings nor the cost reductions that the building's developers had hoped for. Having said that, the marketing appeal of green buildings remains strong, and increased government incentives for investment in renewable energy suggests that these experiments will continue to gain pace.

COGENERATION

Cogeneration (also known as "combined heat and power") involves the recapture of heat from exhaust gases produced during electricity generation. Instead of allowing that "waste heat" to be vented into the environment, it is recovered and used to provide heating or cooling in a building—dramatically increasing the efficiency of the original combustion. Cogeneration tends to work well in large buildings, where the waste heat produced can be used very close to the source of the power. It is generally called on to provide power during peak demand times or when backup power is needed during a utility outage.

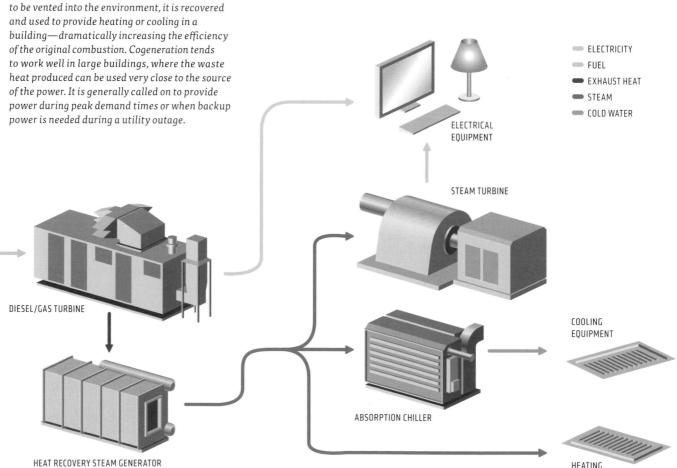

ELECTRICITY
FUEL
EXHAUST HEAT
STEAM
COLD WATER

ELECTRICAL EQUIPMENT

STEAM TURBINE

DIESEL/GAS TURBINE

HEAT RECOVERY STEAM GENERATOR

ABSORPTION CHILLER

COOLING EQUIPMENT

HEATING EQUIPMENT

SOLAR ENERGY

As someone once said, relying on solar energy to power a skyscraper is akin to trying to get a suntan standing up: there isn't much surface area to brown. Nevertheless, attempts have been made to integrate photovoltaics into the skin of skyscrapers. The first major application in the United States was at 4 Times Square in New York City, a building completed by the Durst Organization in 2000. Panels were placed between the thirty-seventh and forty-third floors on the building's south and east façades and remain there today—but have generated only a fraction of the amount of energy it was hoped they would provide.

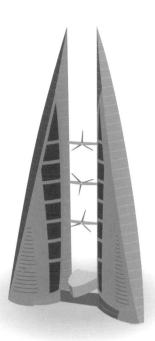

WIND ENERGY

Because winds are stronger at higher altitudes, attempts are now being made to shape buildings in a way that funnels surrounding winds into a zone containing wind turbines. The Bahrain World Trade Center boasts three such turbines, which are set at the center of the building along separate sky bridges that link the two sides of the building. Each turbine measures 95 feet (29 meters) across and is aligned to the north—the direction of the prevailing winds from the Persian Gulf. The developer of the building expects them to operate 50 percent of the time on an average day and to produce between 10 and 15 percent of the building's energy.

FUEL CELLS

A fuel cell is an electromagnetic device that combines hydrogen and oxygen to produce electricity, heat, and water. With no combustion, fuel cells make less noise and emit far less carbon dioxide than traditional engines. However, they must run continuously to be effective, need to be replaced every five to 10 years, and must be serviced more frequently. The largest fuel cell installation in the world is currently planned for 1 World Trade Center (formerly known as the Freedom Tower) now under construction in lower Manhattan.

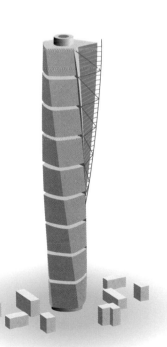

GEOTHERMAL ENERGY

The use of geothermal energy for power dates back decades, but it is only recently that the earth's heat has been used to help power a particular building. By digging deep into the earth's crust (modern drilling can reach down six miles), pressurized steam is released and piped to a power plant, where it produces energy. Excess steam is then collected and reinjected into the underground reservoir. Geothermal technology is commonly employed in parts of northern Europe, including Sweden, where it supports Santiago Calatrava's Turning Torso and other residential buildings in Malmo.

Water Conservation

Like many other green initiatives, water conservation technologies have a "back to the past" ring to them. For thousands of years rainwater has served as a principal source of domestic water for societies around the globe. The cisterns that can now be found on the roofs of skyscrapers in Singapore and across much of northern continental Europe are in many ways a modern-day version of catchment systems devised by the Romans thousands of years ago.

Current attempts to preserve and reuse rainwater are not a response to shortages of clean water, nor are they primarily a cost-saving measure. Instead, they are generally aimed at preventing excess storm water runoff from high-rise towers, which can push municipal sewage systems beyond capacity and lead to the overflow of untreated sewage into local waters. Each inch of rainfall translates into 600 gallons

(2,270 liters) of water per thousand square feet (93 square meters) of roof—water that is better directed toward toilets and other nonpotable building uses.

In addition to rainwater harvesting, several new residential skyscrapers are experimenting with "gray-water recycling" within a building. Water from baths, showers, and sinks is treated to remove contaminants and then returned to serve building needs: toilet and urinal flushing, cooling tower water, general cleaning, and garden irrigation.

Regardless of whether the water is harvested from the sky or recycled, water must be filtered and passed through an ultraviolet treatment before it can be used again. As a result, gray-water technologies require the installation of a dual plumbing system within a building—and are therefore found almost exclusively in newly constructed towers.

Even more experimental are "black water technologies." These aim to recycle, filter, and aerate water from toilets, which typically contains significantly higher levels of nitrogen and organic materials than gray water. A handful of black-water recycling experiments are under way in buildings around the world (with the water used to restock toilets or water plants), but in most places the cost of these systems remains high compared to traditional wastewater disposal options.

Other techniques to conserve water, long popular in Europe but only now finding acceptance on a more global scale, include low-flow plumbing and dual-flush toilets; the latter offers users a choice of using either 0.6 or 1.2 gallons (2.3 or 4.6 liters) of water per flush. Waterless urinals have likewise grown in popularity among designers of green buildings.

Waterless urinals

Urine

Sealant liquid

To the drain

Waterless urinals are increasingly appearing as a water conservation feature in green skyscrapers. Because they do not rely on water supply systems for flushing, they can save millions of gallons of water a year in a large building.

Waterless urinals rely on the "vertical trap" principle for their success: urine initially passes through a floating layer of liquid, which forms a barrier and prevents odors from filling the restroom. After passing through the barrier and into a holding area, the urine flows into a central tube and then down a conventional drainpipe. The liquid within the urinal must be replenished every so often; likewise, the trap at the base catches sediments found in the urine and must be replaced several times a year.

Introduced in the early 1990s, waterless urinals are currently in use in places like the San Diego Zoo, the Rose Bowl, and many civic offices. However, their introduction in skyscrapers has not been without controversy.

When developer Liberty Property Trust announced its intention to install 116 no-flush urinals in its new 57-story Comcast Center in Philadelphia in early 2006, the local plumbers' union was not happy about the reduction in piping (and installation work) that would result and claimed they were "unsafe."

Keen to save 1.6 million gallons (6 million liters of water) a year at one of its newest, greenest towers, the city jumped into the fray. City licensing approval was withheld while a compromise was worked out between the developer and the powerful union. Ultimately, the developer agreed to install traditional supply piping in the walls—to ensure that the toilets could be converted to traditional urinals if the waterless versions did not work.

The Visionaire vision

The Visionaire, a residential building designed by Pelli Clarke Pelli and located in lower Manhattan, boasts one of the most sophisticated water-processing systems in the country. The building's demand of 40,000 gallons (151,000 liters) of water per day is met through a combination of its on-site wastewater treatment system, city-supplied potable water, and collected storm water.

- POTABLE WATER
- STORM WATER HARVESTING
- BLACK WATER TREATMENT
- FILTERED GRAY WATER

The system reduces the building's potable water use by 40 percent.

In addition to the water-recycling system, all residents enjoy rooftop gardens that are designed to collect and filter rainwater while reducing storm water runoff. This water-harvesting system is capable of collecting 5,000 gallons (19,000 liters) per day for toilet flushing and for supply to the building's cooling tower.

WATER MAIN

The Visionaire's innovative wastewater treatment plant treats 25,000 gallons (94,600 liters) of water per day. The building's sewers are linked to an aerated "feed tank" to provide mixing and a "trash trap" to remove all nonbiodegradable solids.

A subsequent biological process, designed to accumulate solids, takes up to 10 hours for water to move through. Liquid/solid separation is accomplished with an "ultrafiltration" membrane.

Treated, filtered effluent then passes through an ozone generation and ultraviolet light (UV) system to remove any traces of color or pathogens. Even after the water is treated and stored, it is continuously circulated through the ozone and UV systems. The system is able to store up to 15,000 gallons (56,800 liters) of treated effluent in its tanks.

Green Materials

Over the past decade the push to make building construction more sustainable has alighted on the idea of using "green materials"—an imprecise term for a range of building-related products. Sometimes used to refer strictly to renewable, reusable, or recyclable resources (such as lumber from managed forests, stone and metal, or recycled steel), it is also used to refer more broadly to a variety of other products—including those with low toxicity, minimal chemical emissions, or high levels of moisture resistance.

Often, "green-ness" is determined not by what the product contains but by the resource intensiveness of its manufacturing or the maintenance it requires once in place. Products such as ceiling and floor tiles, toilet partitions, and wallboard are increasingly evaluated for their "sustainability," as defined on a number of these dimensions.

Some types of products are more easily certified as green than others. Organizations like the Forest Stewardship Council (FSC) and the Programme for the Endorsement of Forest Certification (PEFC), which respectively promote responsible management of the world's forests by certifying forest products in the United States and Europe, have succeeded in creating voluntary, market-driven standards for wood products.

A green room

STEEL
The steel industry has been actively recycling for more than 150 years, because it is cheaper than mining ore. In 2008, more than 97 percent of structural steel was recycled.

GYPSUM BOARD
Gypsum can be recycled and reused, and recycled paper can be used on paper facing. Gypsum board emits virtually no volatile organic compounds (VOC).

CEILING TILES
Ceiling tiles are generally low in VOC but are porous and may absorb emissions from other building materials and reemit them into the indoor air.

CARPETING
"Green labels" show which carpets have been tested for lower levels of VOC by industry and trade groups.

PAINT
Low-VOC paints do not contain harmful solvents that get released into the air as the paint dries, which can cause symptoms such as headaches and dizziness.

WOOD
Green building typically involves wood products that are sustainable, such as bamboo, or have been grown in sustainably managed forests.

Fly ash concrete

Fly ash is a "pozzolana"—named after an Italian city, Pozzuoli, which is regarded as the birthplace of ash concrete technologies. When coal is burned in a power plant it produces microscopic, glassy spheres that are rich in minerals. These particles are collected from the power plant's exhaust before they can "fly" away and are used to produce concrete.

Because fly ash particles are small, they effectively fill voids in the concrete mixture. This makes concrete stronger and more durable and allows it to be produced using less water.

POZZUOLI

Face-lifts for the ladies

Both the Willis (Sears) Tower in Chicago and the Empire State Building in New York are undergoing major renovation programs in an attempt to reduce their energy usage and reliance on fossil fuels.

Willis (Sears) Tower has 16,000 single-pane windows. A window replacement and glazing program could save up to 50 percent of its heating energy.

Water savings through upgrades to fixtures, condensation recovery systems, and water-efficient landscaping will save 24 million gallons (91 million liters) of water each year.

Renewable energy technologies, like wind and solar, and innovations such as green roofs will be tested to take advantage of the tower's height and unique setback roof areas.

Mechanical system upgrades will take the form of new gas boilers that utilize fuel-cell technologies as well as new high-efficiency chillers.

Lighting power density is reduced in tenant spaces by using ambient and task lighting, dimmable ballasts, and photosensors that dim lights according to daylight availability.

Approximately 6,500 double-hung windows will be upgraded with suspended, coated film and gas fill. "Remanufacturing" of the window units will take place within the building itself.

More than 6,000 insulated reflective barriers will be installed behind perimeter radiator units.

A chiller plant retrofit will include four industrial electric chillers in addition to upgrades to controls and variable-speed drives.

WILLIS (SEARS) TOWER   **THE EMPIRE STATE BUILDING**

Retrofitting

Despite the spate of tall buildings completed recently in Asia and the Middle East, the fact remains that most skyscrapers are old. Over three quarters of all buildings in the United States are more than 20 years old—and the same is true for high-rise towers. As a result, many are undergoing face-lifts to integrate sustainable, or "green," technologies that simply did not exist when they were built.

Old buildings are not necessarily energy inefficient. Because of their bulk and thermal mass, most buildings completed before 1950 have energy requirements far lower than the glass curtain wall structures built subsequently. Nevertheless, competitive pressures have required their owners to improve lighting levels, enhance existing glazing, and introduce better space conditioning. The significant investment required to bring them up to today's standards will hopefully see a return—both in lower operating costs and in higher rents.

Two of the tallest and most famous buildings in the United States, the Empire State Building and the Willis (Sears) Tower, initiated major retrofit programs in 2009. The Empire State Building's program, involving expenditure of roughly $20 million, is expected to save $4.4 million annually in energy costs. The Willis (Sears) Tower's five-year program is forecast to cut electricity use by roughly 80 percent and save 24 million gallons (91 million liters) of water each year.

Measuring Sustainability

Around the world, a variety of voluntary systems have been promulgated to assess the sustainability of both new and existing buildings. In the United States, the United States Green Building Council was founded in 1993 as a nonprofit entity to promote sustainable construction. Its LEED (Leadership in Energy and Environmental Design) environmental impact-rating system debuted in 2000 with six parameters: sustainability of the site, water efficiency, energy and atmosphere, materials and resources, indoor air quality, and innovation/design processes.

Four award levels, based on a point system, were established: certified, silver, gold, and platinum. Just how significant the additional cost of achieving these levels is as a component of a building's cost is unclear, but it is generally thought to range anywhere between 1 and 10 percent. Nevertheless, the number of buildings applying for LEED certification has increased dramatically: in 2003,

only 84 buildings were certified, but by 2008 the number had grown to over 1,400.

Reliance on the LEED program in the United States will increase further over the next decade as many of the largest cities in the country—including Boston, Los Angeles, and San Francisco—have mandated LEED certification as a prerequisite for receiving approvals for new buildings. Some places are even offering tax incentives for sustainable upgrades. Cincinnati has offered a property tax rebate on LEED-certified buildings of $500,000 annually for 15 years on new buildings and 10 years on existing ones; the state of Nevada has likewise established incentives.

Nearly all European countries have established systems similar to LEED—indeed, some are even older than the American system. BREEAM (Building Research Establishment Environmental Assessment Method) was set up in 1990 in Britain to measure the sustainability of commercial buildings—including schools,

retail stores, hospitals, and factories. Similar systems are now in operation in The Netherlands, France, and Spain. Farther afield, Australia, South Africa, and New Zealand rely on a system known as "Green Star."

Unfortunately, there is not necessarily a direct correlation between the level of certification a building achieves and its actual energy performance. Indeed, early studies have shown that LEED-rated buildings, from an energy-efficiency standpoint, often underperform more traditional "uncertified" buildings.

Nevertheless, there is clearly some value to be extracted from participating in the sustainability ratings race. Recent surveys have shown that LEED status confers competitive advantages in the real estate marketplace; it was recently reported that LEED-rated buildings in New York had higher occupancy (by 4 percent) and commanded higher rents ($11 per square foot more) than nonrated buildings.

Green labels

The appeal of building green is universal. While there are four predominant rating systems in the world—LEED, CASBEE, BREEAM, and Green Star—the way each achieves its goal of sustainability differs based on its native country's climate and culture.

- ■ BREEAM
- ■ CASBEE
- ■ GREEN STAR
- ■ LEED
- ■ OTHERS
- ■ EMERGING
- ■ POTENTIAL INTEREST

America's greenest skyscraper?

One of the newest skyscrapers in New York, One Bryant Park (also known as the Bank of America Tower), is also its most sustainable. Located on the corner of West 42nd St. and 6th Ave., its tapered and elegant glass form belies a series of green initiatives that— if successful—may set the standard for American skyscrapers of the future. Soon after its opening in 2009, its developers, the Durst Organization, received "LEED Platinum" status for the building from the U.S. Green Building Council—making it arguably the greenest skyscraper in America.

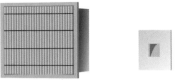

Roughly 95 percent of the dust, dirt, and particulates are filtered from the air that enters the Bank of America Tower. It is said that the exhaust air, on leaving the building, will actually be cleaner than the outside air.

Over a third of all materials used in construction were recycled and 85 percent of the resulting construction debris was recycled. More than 20 percent of construction materials came from within a 500-mile radius of the site.

The tower is enclosed in specially coated glass with fritting that allows visible light to pass through while reflecting ultraviolet rays, reducing both lighting and cooling loads.

Specially designed chutes on every floor allow tenants to dispose easily of a variety of recyclable materials.

A geothermal system is used to cool the building in the summer and heat it in the winter. Wastewater from this system is collected with storm water runoff and used for toilet-flushing water.

At night the cogeneration plant makes ice in the basement. As the ice melts the next day, cold water is piped to the air-conditioning system to reduce the building's peak energy demand.

The tower houses a 5.1 megawatt cogeneration plant. It captures the heat it generates and puts it back to work, converting it to energy used to heat and cool the building.

DREAMING IT

THE FUTURE

Less than a century ago architects expressed their visions of an urban future based on the new skyscraper typology. Le Corbusier was among the earliest, with his 1923 proposal for the Ville Contemporaine—a series of 60-story glass-clad office buildings set in open parkland above a transportation hub— and his subsequent proposal for the Ville Radieuse (or Radiant City), which extended the concept further by specifying zones for working, living, and leisure. Though never realized in their purest form, Le Corbusier's radical ideas for reshaping blighted city centers became the foundation for subsequently much maligned high-rise public housing complexes in Western Europe and the United States.

Even more fanciful were some of the ideas of the American architect Hugh Ferriss. In the 1920s he imagined skyscrapers that would do more than serve as commercial trophies: they would provide homes to people, be integrated with new forms of transit, incorporate retail and cultural areas, and provide outdoor terraces at great heights. Though he would not live to see it, by the end of the century many of Ferriss's novel ideas for skyscraper life

had become common features of high-rise developments around the world.

Another celebrated tall-building visionary was Frank Lloyd Wright, who promoted his own revolutionary idea for the skyscraper of the future in 1956. Wright's proposed Illinois Tower reached a mile into the sky. Its 528 floors and 18.5 million square feet (1.7 million square meters) were intended to serve 100,000 office workers. While the idea was a physical impossibility then, and remains highly impractical today, Wright's vision became a touchstone of sorts in the ongoing race to the sky.

The twenty-first century has brought its own visions of the future—both derivative and unique. The nearly mile-high Nakheel Tower proposed for Dubai replicates in many ways the features of the emirate's recently opened Burj Khalifa, only taller: it would stretch to 4,600 feet (1,400 meters), nearly 2,000 feet (610 meters) taller than its neighbor. A proposal to build a "Mile-High Tower" near Jeddah, Saudi Arabia, echoes Wright's own scheme. The Burj Mubarak al-Kabir in Kuwait is proposed to reach 3,300 feet, or 1,001 meters—a nod to the "Arabian Nights" collection of stories.

None of these projects are likely to move forward quickly (if they move forward at all), given the current financial climate and the time it could take for the global economy to recover. Nonetheless, the

proposals are emblematic of certain trends in the evolution of skyscrapers—trends already defining the nature of our twenty-first-century cities.

Most notable, perhaps, is the fact that skyscrapers are no longer an American phenomenon. Like other American inventions that helped define the modern age but were perfected by Asian companies (such as the television, the car, and the radio), the skyscraper is today, in its most aggressive form, largely an Asian phenomenon. Hong Kong has far more buildings over 300 feet (90 meters) in

Architect Hugh Ferriss imagined skyscrapers serving many purposes, including holding up bridges.

height than New York. The Petronas Towers, in Kuala Lumpur, and Taipei 101, in Taiwan, each held the title of world's tallest building. And the three new towers in Shanghai's Pudong district (Jin Mao, the Shanghai World Financial Center, and the future Shanghai Tower) comprise the greatest concentration of supertall buildings anywhere in the world today.

Building tall has spread well beyond Asia. Mirroring recent changes in the global economy, the Middle East has also embraced the new urban form wholeheartedly. Hundreds of residential and mixed-use towers have sprung up across the Middle East over the last decade, many incorporating designs and technologies as yet untested in the United States. When the Burj Khalifa, the world's tallest building, opened to great fanfare in Dubai in early 2010, it was not alone: 14 other supertall buildings opened in the Middle East between 2000 and 2010 as well.

American influence is still very much apparent in skyscrapers and skyscraper-related technology. Pick any supertall building under construction in Asia or the Middle East and chances are that the architect, structural engineer, or foundation consultant will be American. But the money and the hubris that fuel these multibillion-dollar construction projects abroad are distinctly local—and are likely to remain so for the foreseeable future.

Le Corbusier's urban vision centered on a series of high-rise towers set in open parkland.

A mile-high tower

In the mid-twentieth century, Frank Lloyd Wright publicized his idea for a mile-high tower, the Illinois. Wright imagined it the visual centerpiece of "Broadacre City," a community of low-rise homes, each on an acre of land. In many ways his tower had more in common with the church spires of traditional villages than it did with notions of an urban skyscraper.

Wright's Illinois had 528 floors, with a spire that reached 5,280 feet (1,600 meters) into the sky. To move its 100,000 inhabitants up and down, it relied on five-story elevators running on ratchet interfaces located on the outside of the building (to conserve space in the interior). In total, it covered 18.5 million square feet of space (1.8 million square meters)—almost seven times the size of the Empire State Building.

Though certain aspects of the proposal were intriguing, Wright's tower could not possibly have been constructed in 1956. All-concrete buildings typically rose no more than 20 stories then. Today advances in concrete technology have resulted in a six-fold increase in concrete's strength under compression, allowing it to underpin buildings as large as Shanghai's Jin Mao, at 88 stories.

Though Wright's tower probably could be built today, it would likely be uninhabitable. Moving people up and down the tower in reasonable time would require elevator speeds well beyond the level of human comfort. Enormous amounts of money and materials would have to be spent reducing sway. And once the lateral load-resisting structure and the number of elevators needed to serve such a large population were in place, there would be little room left for living or working. In other words, the mile-high tower—as august as it sounds—makes no economic or commercial sense.

How High?

How high the next generation of buildings will go is unclear. Dubai's Burj Khalifa reaches 2,717 feet (828 meters)—a full 60 percent taller than Taipei 101, the previous holder of the height record. Given its size, the world economic climate, and the lead time involved in constructing a supertall building, the Burj is likely to remain the tallest building on earth for at least the next decade.

A number of other supertall towers will be completed in the next few years. The 121-story Shanghai Tower, currently under construction, will take its place in 2014 alongside the Jin Mao Building and the Shanghai Financial Center in Shanghai's Pudong district. Also targeted for completion in 2014 is the Lotte Super Tower in Seoul, Korea. It will reach 1,821 feet (554 meters) into the sky, making it Asia's tallest building.

In London, construction has commenced on The Shard, a mixed-use complex adjacent to London Bridge Station that will be Europe's tallest skyscraper. In New York, construction is under way on another local record setter; completion of 1 World Trade Center (formerly known as the Freedom Tower), with an antenna reaching to 1,776 feet (541 meters), is slated for late 2013.

Several supertall projects have fallen victim to the economic climate. The Calatrava-designed Chicago Spire, which would have made Chicago home once again to America's tallest building, was put on hold in 2008—after the hole for its foundation was dug. Plans for the 200-story Nakheel Tower in Dubai were suspended indefinitely. Construction of the Gazprom Tower in St. Petersburg was put on hold in mid-2009 and a decision to relocate it made late in 2010.

Like other fanciful buildings, postponed projects like Nakheel may never get built. Supertall buildings are extremely complicated to design, require a very robust leasing and sales market, and take more time to construct than most lenders are willing to accept. Rarely do they incorporate the sort of engineering efficiencies needed to make a development project commercially viable—which means that, absent support from a royal family or a flush private corporation, they simply won't happen.

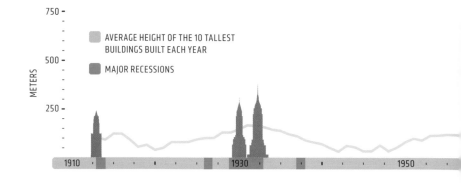

AVERAGE HEIGHT OF THE 10 TALLEST BUILDINGS BUILT EACH YEAR

MAJOR RECESSIONS

Building tall before a fall

There is a correlation between the timing of record-setting skyscrapers and the economy. The construction of "the world's tallest building" has historically been initiated toward the end of a boom period, when demand appears strong, debt financing is attractive, and high land prices drive up the number of stories needed to offset land costs. These record setters are rarely finished before a recession has begun and tenants are scarce.

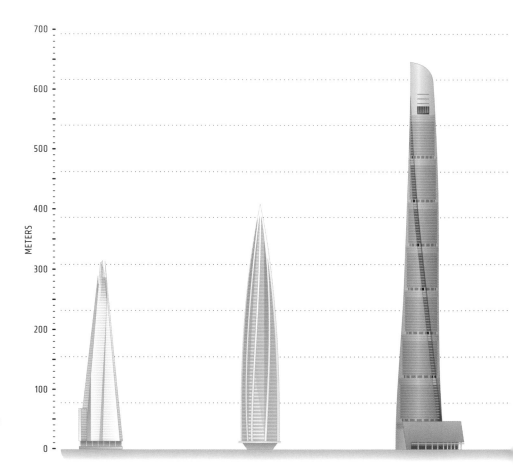

The Shard at London Bridge, designed by Renzo Piano and funded largely by Qatari investors, will become Europe's tallest tower when it opens in 2012.

The five sides of the Gazprom Tower, originally slated to rise in St. Petersburg, will twist as they reach the sky—if and when it is built.

The Shanghai Tower will be the world's first double-skin supertall building, with the world's highest nonenclosed observation deck, when it opens in 2014.

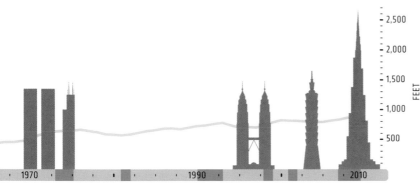

- 2,500
- 2,000
- 1,500
- 1,000
- 500

FEET

1970   1990   2010

Such was certainly the case in 1913, with the construction of the Woolworth Building , and again in the late 1920s and early 1970s. The Empire State Building (1931) remained nearly empty for a decade, and the World Trade Center (1973)

might have as well had the state government not relocated there. And the Burj Dubai was renamed Burj Khalifa just before its 2010 opening—after the emirate was bailed out by neighboring Abu Dhabi (Khalifa is the United Arab Emirates' president).

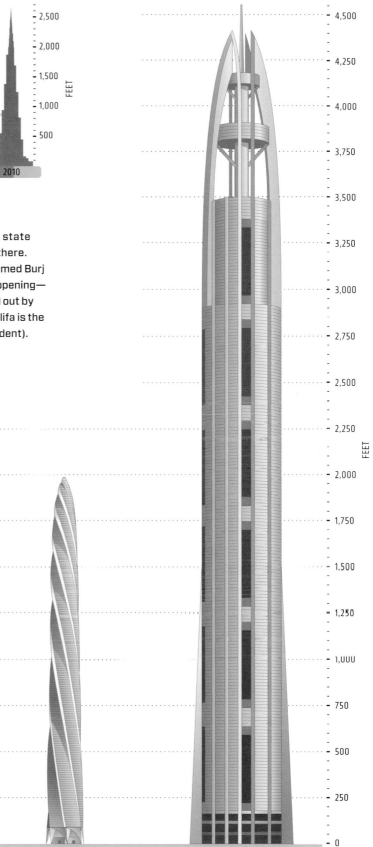

- 4,500
- 4,250
- 4,000
- 3,750
- 3,500
- 3,250
- 3,000
- 2,750
- 2,500
- 2,250
- 2,000
- 1,750
- 1,500
- 1,250
- 1,000
- 750
- 500
- 250
- 0

FEET

The 105 floors of 1 World Trade Center, under construction in New York, will be capped by a television antenna reaching a symbolic 1,776 feet into the sky.

The 123 floors of the Lotte Tower in Seoul will contain retail, offices, residences, a luxury hotel, and an observation deck.

Though ground was broken and foundations laid for Santiago Calatrava's Chicago Spire in 2007, the project was put on hold in late 2008.

Construction of Dubai's proposed Nakheel Tower—four separate towers linked at 25 levels by sky bridges—would likely require a decade to complete.

How Green?

Just how sustainable a high-rise tower sheathed in glass can ever be is a subject of ongoing debate among architects and engineers. Certainly the amount of embodied energy and carbon emissions involved in constructing a building at great height will never be offset by environmental credits the building might amass later in life, nor will the energy produced on-site come close to covering the energy demands of the building when tenanted.

Yet new skyscrapers in dense urban areas are, by dint of their location, generally greener than other types of commercial and residential buildings. They are typically located near mass transit, minimizing the fossil fuels consumed by private cars

and the negative environmental impacts associated with driving. Vertical living also requires less energy for heat: city dwellers take up less space and use less energy per capita than suburban or rural residents. And these new buildings are designed to last—up to a century or more.

Nevertheless, designers of skyscrapers around the world will continue to go to great lengths to minimize the environmental footprint of new towers. These efforts will take many forms: orienting the building better to the sun and the wind, expanding the use of natural light and ventilation, providing more sophisticated thermal barriers in curtain wall design, maximizing the

use of renewable energy (both solar and wind), ensuring better collection and utilization of rainwater, and conserving energy through intelligent building management systems.

In some countries sustainable building practices are now mandated by law. As a result of a European Union directive, buildings completed from January 2019 in EU countries will be asked to produce as much energy on-site as they consume—one definition of what is loosely referred to as a "net-zero energy building." Member states have also been asked to establish two sets of national net energy reduction targets for existing building stock, to be achieved by 2015 and 2020, respectively.

The 71-floor Pearl River Tower, a new corporate headquarters for the Chinese National Tobacco Company, is nearing completion in Guangzhou. Its design was driven by the desire to minimize carbon emissions through a variety of features, including photovoltaics, wind turbines, and a high-performance skin.

The Marina Bay Sands in Singapore is one of the largest hotel-convention center-casinos in the world, and likely the greenest. Completed in 2010, its three towers support a three-acre (1.2 hectares) "sky park" on the roof, which was lifted into place (55 floors above the city) in sections after the towers were completed. At one end the park cantilevers out 200 feet (60 meters) beyond the tower's edge—roughly the length of a 747 airplane.

Harnessing natural forces, Adrian Smith+Gordon Gill's proposal for the Clean Technology Tower in Chicago places wind turbines at the corners of the building, to capture wind at its highest velocity, and includes a domed double-roof cavity to direct wind toward an array of turbines used for natural ventilation. The dome itself would be shaded by solar cells that capture the southern sun.

The proposed Rodovre Sky Village in Copenhagen is more than a mixed-use building intended to house both office and residential uses. The tower would consist of stackable green-roofed units, which can serve as office or home (or parking spaces) and whose internal spatial layout offers maximum flexibility to future users of the space.

In the United States, the concept of a net-zero or zero-energy building remains elusive. To date there is no agreement on how such a building is defined—whether by its energy cost, its emissions, the energy used on-site, or the total amount of "source energy" (including the energy used to make the energy consumed). And while the absence of a definition has not stopped the federal government from setting long-term net-zero targets for commercial buildings (nor has it stopped states like Massachusetts and California from passing their own energy consumption laws), it remains to be seen how implementable or achievable any of these targets will be in practice.

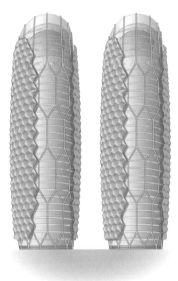

The design concept of the ADIC Headquarters, now under construction in Abu Dhabi, derives from traditional Islamic patterns. Conceived as a dynamic façade, the mashrabiya *will open and close in response to the desert sun's path—protecting the most severely exposed parts of the building and contributing to a projected 25 percent reduction in the total cooling load.*

The "bioclimatic" design proposed for the EDITT Tower in Singapore features vegetation that spirals from street level upward. Covering half the building, this would form a continuous ecosystem and facilitates ambient cooling of the façade. A rainwater-collection system on the roof and along the façade would direct water through a gravity-fed, water-purification system for reuse.

Vertical farming?

Some of the most outlandish proposals for future skyscrapers have revolved around the concept of "vertical farming." By moving agriculture indoors and relying on hydroponic and aeroponic techniques, one acre (0.4 hectare) of indoor farm can theoretically replace four acres (1.6 hectares) of farmed land outdoors. A 30-story building, it is estimated, could feed 50,000 people.

To understand farm math, it is necessary to quantify the waste and environmental damage associated with traditional farming. These "inefficiencies" range from weather-related crop failures to pesticides and agricultural runoff to the emissions resulting from farm equipment like tractors and plows. Vertical farming would, in theory, remove inefficiencies such as these.

The premise relies on a number of rather far-reaching assumptions, including the location of vertical farms at transit hubs along arterials—so that food can be moved to consumer destinations easily. It also assumes sophisticated plant technology, including smart temperature controls that maintain precisely the right growing environment and a "gas chromatograph" that analyzes the "flavinoids" of the plant and indicates precisely the right moment to harvest.

What Shape?

Not all tall buildings coming online in the short-term future will be supertall. Most are likely to distinguish themselves in other ways—by the materials used as their skin, by their green credentials, or by their shape.

The shape of today's skyscrapers is particularly notable. The advances in technology and materials that have allowed buildings to reach 160 stories have also allowed them to take on new and exciting shapes. Today tall towers can twist, lean, and turn back on themselves in ways that would not have been possible 50 years ago. These radical shapes are generally chosen for dramatic effect, but occasionally they contribute to minimizing wind loads by improving a building's aerodynamic properties.

Radical shapes are not limited to any type of building or part of the world. Residential buildings, such as the Puerta de Europa in Madrid or the Turning Torso in Malmo, appear to lean precariously. Office buildings such as the CCTV headquarters in Beijing or the Swiss Re Building in London function proudly as corporate headquarters despite their untraditional shapes. And some complex designs, like the sweeping façade of the proposed Empire Island Tower in the Middle East, have been chosen for mixed-use developments serving hotel, office, and residential users.

TRUMP INTERNATIONAL HOTEL AND TOWER

The proposed design for the Trump Organization's first venture in the Middle East features a split but connected 62-floor tower, so designed to minimize shadows.

SWISS RE

The unusual shape of Foster+Partners's London office tower, fondly known as "the Gherkin" since opening in 2004, causes air to move over its surface in ways that assist in ventilating the building.

TURNING TORSO

Based on one of his own sculptures and completed in 2005, Santiago Calatrava's Turning Torso residential building in Sweden is composed of nine units that turn 90 degrees along the tower's 600-feet (183-meter) height.

EMPIRE ISLAND

Located one block from the sea in Abu Dhabi, the proposed Empire Island residential tower sweeps back on itself to maximize views and light.

CRYSTAL CITY

Conceived as a city within a city, Foster+Partners's proposed Crystal Island would enclose 27 million square feet (2.5 million square meters) in a tentlike superstructure at a site close to the center of Moscow.

CCTV

Rem Koolhaas's CCTV Tower in Beijing is the city's tallest. Its three-dimensional "cranked-loop" shape houses radio and television broadcasting studios as well as restaurants and an observation deck.

The next generation

The 100 tallest buildings under construction span the globe, but the number of projects in Asia and the Middle East suggests that the United States will no longer be home to the majority of the world's tallest buildings.

WORLD'S 100 TALLEST BUILDINGS TO BE
COMPLETED BY 2013 (IN NUMBER OF BUILDINGS)

1   5   15   30

In the United States, several supertall skyscrapers have been designed and permitted in Chicago and New York, but the recession has undercut their financial viability, and development work has been halted in many places.

In the Middle East, tall buildings continue to rise in Qatar, Saudi Arabia, Kuwait, and Dubai both for residential and commercial use—though none threaten to rob the Burj Khalifa of its world's tallest status.

In Asia, China, Hong Kong, and Singapore remain home to active skyscraper construction but have been joined by Korea and India—both of which have record-breaking projects under way.

Where?

The shift away from the United States as a home to the world's tallest buildings has been dramatic and quick. In 1980, 49 of the top 100 of the world's tallest buildings were located in Chicago and New York— including nine of the top 10. By 2010 this number had dropped to 18. Today, of the world's top 10 tallest buildings, only the Trump International Hotel and Tower and the Willis (Sears) Tower—both in Chicago—are in the United States.

There is no sign that this geographic trend will reverse itself anytime soon. According to the Council on Tall Buildings and Urban Habitat, of the 100 proposed

future tallest buildings only five of them would be based in the United States. Nearly two thirds would be located in Asia and more than 20 of them would be in the Middle East. Notwithstanding the fact that many of these will never be built, the number of skyscrapers proposed for the Middle East and Asia indicate an enthusiasm for size that has largely been absent in the United States since the recession following the attack on the World Trade Center in 2001.

Supertalls aside, the construction of "regular" tall skyscrapers has continued over a wider base than ever before. Three quarters of the buildings over 650 feet (200

meters) completed in 2009 were in Asia and the Middle East; only one quarter of them were located in North America. They were spread over 25 cities (nine of them in China)—a far cry from the days when only Chicago and New York embraced the form.

This explosion in tall buildings outside of the United States looks set to continue into the next decade. Indeed, more skyscrapers were under construction at the beginning of 2010 (370) than were built in the entire last decade—and in more places. China remains the most active market, but India and Korea now boast a dizzying array of skyscraper construction sites as well.

How Will We Live?

The skyscraper of the future will bear little relation to the ones of the past. No longer will it serve solely commercial purposes; instead, it is as likely to be a mixed-use destination—including retail stores, commercial offices, and hotel or residential accommodation. It may also house restaurants and recreational space, supermarkets, movie theaters, swimming pools, or libraries.

The increasing popularity of mixed-use complexes, and in particular the growth in residential towers, has left its mark on every aspect of skyscraper design and construction. In terms of structure, concrete has now overtaken steel as the most prevalent skyscraper material. Representing only 10 percent of the core structural systems of skyscrapers over 650 feet (200 meters) in 1970, concrete today is responsible for a full 40 percent—with another 30 percent of new towers comprised of composite (steel and concrete) beams and columns.

In terms of construction, these mixed-use buildings are far more difficult and costly to erect than single-purpose ones. Often, a concrete residential tower will rise from a steel-framed office podium, increasing the number of trades on the job and making more complex the already delicate job of choreographing construction staging. Add to that the mechanical systems that must support multiple elevator banks, ventilation and plumbing systems, and restaurant facilities, and you have one very complicated construction job.

In terms of design, these mixed-use buildings present the added complexity of segregating users and uses—in terms of pedestrian flows, vertical transportation, noise attenuation, loading, and other services. In designing these buildings, architects must often deal with multiple building code provisions, as standards for commercial and residential occupancy (let alone observation decks and concert halls) often differ.

It is here, in the design of today's mixed-use towers, that one finds some hint as to what skyscraper life might be like in the future. Though almost all are privately owned, parts of many mixed-use buildings are today designed to function as a form of public space: witness, for example, the town

City in the sky

The Burj Khalifa, formerly known as the Burj Dubai, is the first mixed-use building to hold the title of world's tallest building. Formally opened in early 2010, it rises to 2,717 feet (828 meters) and contains 160 inhabitable stories.

An exclusive restaurant is located on the one hundred twenty-second floor of the building.

Several spas are located in the building, including a four-story fitness and recreation annex.

Both a library and a cigar club are located within the complex.

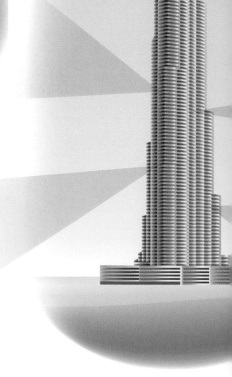

Roughly 220,000 square feet (20,400 square meters) of offices can be found at the top of the tower, on floors 123 through 160. A lounge lobby on the one hundred twenty-third floor welcomes workers and guests.

The building contains over 1,000 residential units. "Wind tracker" information is available in the apartments with terraces so that residents can be alerted to wind conditions outside. Liquid crystal display panels in the hotel rooms (and in the apartments) convey information in an emergency.

Four pools are located in the complex.

An Armani-branded hotel, with 160 guest rooms, is located between the fifth and eighth floors of the Burj, with additional suites on floors 38 and 39. The hotel includes eight restaurants, a 12,000-square-feet (1,100 square meters) spa, and a florist—all open to the public.

Retail offerings within the complex include a supermarket.

square-like feel of the retail floors of the mixed-use Time Warner Center in New York at lunchtime almost any day. If historically it was the street or square where the public met to socialize, today it is increasingly the public portions of mixed-use complexes that act as meeting points for residents, office workers, shoppers, and tourists.

Not all attempts at social mixing depend on transient visitors, and not all happen at or near ground level. Internal sky courts are increasingly appearing on buildings in Asia, driven primarily by the requirement for refuge floors. The new Shanghai Tower, for example, will feature vertical neighborhoods separated by "sky gardens"—intended to combine the best of the indoors and the outdoors in a communal gathering place. Some commentators see these sky courts as a vehicle of social integration, akin to the city's historic courtyards or the commercial arcades of the nineteenth century—though just how much social mixing will ever occur in private spaces high up in the sky is unclear.

It is hard to separate the future of the new mixed-use skyscrapers from their impact on the cities around them, and one wonders what someone like Jane Jacobs, one of the great urbanists of the twentieth century, would have had to say about them. In her seminal 1961 work, *The Death and Life of American Cities*, she railed against the spate of "tower-in-the-park," tall residential buildings that she felt were, rather than revitalizing blighted portions of American cities, actually robbing them of their diversity.

How well the mixed-use tower of today meets the criteria Jacobs established for a diverse and lively neighborhood is one standard by which we can measure it. Jacobs would have liked the way the combination of residents and businesses operates around the clock but might have bemoaned the fact that users are not always using common facilities. And she would have approved of the density that a mixed-use skyscraper brings. But she would likely have worried about the homogeneity of class and economic status that an all-new building imposes: without subsidies or regulation it is unlikely to feature low-rent units that bring diversity of both commerce and people.

The long-term viability of the mixed-use tower as an urban form can also be assessed by measuring its "sustainability." From an environmental perspective, vertical living makes sense. Energy use in a sprawling suburban home far exceeds that of a city apartment, notwithstanding the fact that today's glass curtain wall is inferior—in terms of solar and thermal gain—to the traditional masonry façades that preceded it. And city apartment dwellers typically rely on their automobiles far less than residents of suburban or rural areas. Today's move to "transit-oriented development" means that more and more high-rises are being built at or adjacent to transit—further reducing the amount of roadway traffic and sidewalk congestion normally associated with dense urban cores.

The question of sustainability of form, however, is more than an environmental one: it is equally an economic and social one. From an economic perspective, the mixed-use typology makes commercial sense. The spaces at lower levels of a mixed-use complex are used for retail and office functions, which require bigger floor plates and easy access from the street or parking; those at the higher levels are reserved for residential and hotel uses, which demand smaller floor plates and can monetize the views. Operating around the clock, they foster an intensity of use and volume of foot traffic that is conducive to successful and stable retail.

Above all, however, it is the social or sociological sustainability of the mixed-use skyscraper that presents perhaps the hardest questions to answer. Just how well do these "vertical cities" relate to the broader urban fabric? Are these the "urban neighborhoods" of the future or will they become islands onto themselves? Can private sky gardens or elevators ever serve the social mixing purpose that public parks or city streets do? Will mixed-use towers truly lead to more vibrant around-the-clock urban cores or instead to further social stratification? To what extent should municipal planners and governments regulate or incentivize their construction?

It is too early, of course, in the evolution of this new form of skyscraper to know the answers to these questions. What we do know, thanks to the examples of places like Hong Kong, Singapore, and Manhattan, is that vertical living can be wildly successful and need not rob a city of the diversity that makes it great. We also know, thanks to cities like São Paulo, Mexico City, and Los Angeles, that the alternative means of absorbing large numbers of new city dwellers is horizontal sprawl—which inevitably puts unacceptable demands on both local infrastructure and the environment.

The skyscraper as an urban form, then, is likely to grow in popularity—if only to absorb the great numbers of people moving into our cities. But while it is too early in its history to assess the ultimate impact of vertical living on the urban environment, it is never too early for architects and engineers to imagine what skyscrapers could become, just as Frank Lloyd Wright, Le Corbusier, and Hugh Ferriss did almost a century ago.

Today their wildest musings are not only about height—though many proposed buildings do seem to defy the laws of both gravity and real estate economics. The skyscraper visions of the early twenty-first century are as much about absolute size, and more specifically about giving form to the idea of "a city within a city." While the technology to build the most ambitious of these, like the Shimizu Pyramid proposed for Tokyo, does not yet fully exist, the notion of a self-contained city that they celebrate may—like Wright's mile-high tower—set a direction and a challenge for the skyscraper dreamers of the future.

For centuries the tallest buildings in the world, the great cathedrals of Europe, were located at the center of population hubs and reflected the growth of religious power during the late Middle Ages.

The Great Pyramid of Giza, built in the desert outside Cairo in 2500 BC as a monument to a ruler's power, was the world's tallest structure for thousands of years.

Reaching for new heights

Some visions of the future extend the notion of a vertical community by incorporating the idea of a "city within a city." The proposed Shimizu Pyramid would house 750,000 people in a floating community connected by 86 miles of horizontal and diagonal tunnels and built on 36 massive piers in Tokyo Bay. The design displays a certain historical symmetry by returning to the pyramidal form that marked the beginning of humanity's race for the sky.

The modern idea of mixing people and commerce to create an around-the-clock vertical community came a century later and has been embraced most fully in the Middle East and Asia—sometimes as a way to create density and encourage economic development.

The first inhabitable skyscrapers were built in the early twentieth century in downtown Chicago and New York—and were solely commercial in nature.

Acknowledgments

ver the course of my writing life, I have been blessed with terrific research and graphic design teams. But none rivals this one in terms of its capability, enthusiasm, and creativity. Working on *The Heights* with George and Rob over the course of the last year has been both a pleasure and a gift.

George Kokkinidis, the graphic designer responsible for the look and feel of the book, patiently and enthusiastically designed every single page of this book with me—and then managed a team of far-flung illustrators who brought our designs to life. A true professional, he did much more than simply translate my ideas into lines and colors: he read through all of the research material himself to make sure that our vision for each and every graphic was as compelling as possible before we commissioned it. His passion for the book's subject comes through on every page.

Rob Vroman, our talented and thoughtful researcher, found facts and references in places we didn't know existed—and always in record time. As a trained engineer, he was far more comfortable with the technical aspects of the book than either George or I was—and we relied on him heavily to explain anything at all complex (e.g., vortex shedding or cathodic protection). Some of his explanations still have yet to sink in: no matter how many times Rob explains it, I fear I will never truly understand what precisely is happening inside an air-handling unit.

The book itself I owe to Liz Van Doren, who approached me while she was at Black Dog Press about producing a book on how skyscrapers work; she had seen a book George and I did for Penguin Press in 2005 (*The Works: Anatomy of a City*). Though Liz did not ultimately have the chance to edit or publish the book, we are grateful for her inspiration and her timing. Had there not been a massive downturn in the economy soon after she contacted me, all of us might have been too busy to undertake this ambitious extracurricular project.

To make certain sections of this book come alive, I have relied freely on examples drawn from my recent tenure at Vornado Realty Trust. For that I owe thanks to Steve Roth and David Greenbaum, who gave me the chance to help design two fabulous commercial towers in Midtown Manhattan—which I hope will one day be built. I am equally grateful to other Vornado colleagues—in particular to Eli Zamek and Sandy Reis, for teaching me what I needed to know to write this book and for reviewing and correcting the first two sections of this book, and to Gaston Silva and Suki Paciorek, for taking the time to make sure that what I have written in the last section is largely accurate.

Many others who I met while at Vornado have helped to shape the book. I am grateful to Bill Pederson, Dennis Austin, and Rafael Pelli for making commercial building design as exciting as any other form of architecture. I have great respect and admiration both for them as individuals and for the work their firms do (Kohn, Pedersen, Fox Associates; Rogers, Stirk, Harbour + Partners; and Pelli Clarke Pelli Architects, respectively). I am likewise indebted to Jeff Smilow at Cantor Seinuk, Marc Colella, and Denzil Gallagher at Buro Happold—who made the less-than-sexy topics of building structure and mechanical systems come alive for a nonengineer.

There are people I did not know but sought out for their expertise, and they gave it generously. Bill Baker at Skidmore, Owings, and Merrrill and David Scott at Arup each showed me, on their desktop computers 1,500 miles apart, how a structural engineer designs tall buildings to withstand wind at great heights. Mel Febish and Clyde Baker shared more than I ever wanted to know about driving piles and caissons in New York and Chicago, respectively. Lance Hosey raised some interesting questions about sustainability—which I hope have been adequately conveyed. And Chief Thomas Jensen of the Fire Department of New York patiently answered dozens of questions about life safety, ranging from the technical to the philosophical (like why building codes differ so dramatically across the globe).

A variety of people provided information for specific illustrations. I am grateful to Amy Marpman of Great Forest, to John Gilbert of Rudin Management, to Ko Makabe at Kohn Pedersen Fox, to Terry Carbaugh at Turner Construction, to Kevin Huntington at Jenkins and Huntington, to Ambrose Aliaga-Kelly at Gensler, to Helena Durst, and to Rich Miller at Con Edison for their assistance with particular topics. And a special debt of thanks goes to Carol Willis, director of New York's Skyscraper Museum and author of a terrific book about skyscrapers—without whom our chronology of the evolution of the skyscraper would likely have been both incomplete and inaccurate.

I am once again grateful to my editor and publisher, Ann Godoff, and to her team at Penguin Press, who displayed as much enthusiasm for this new book as they did for the last. Likewise, thanks are due to Sloan Harris and Kristyn Keene at ICM, whose understanding of the peculiar economics of illustrated books proved crucial to getting this one started.

Last but not least, I want to convey my deep gratitude to two friends at the Rockefeller Foundation—Peter Madonia and Pilar Palacia—whose encouragement and support helped take what could have been a rather ordinary book and turn it into what we hope is an inspired one. It is rare for authors and designers to derive such immense pleasure from the act of composing a book; even more rare is the opportunity to do it in such a magical place.

Contributing Artists

Photography credits

Index

Airship to Nowhere

When the Empire State Building's developers decided the original 85-floor design "needed a hat," they made it a useful one: a spire that would serve as the western docking station for airships (blimps) coming from Europe. Winch equipment was installed, the building's framework stiffened, and the eighty-sixth floor fitted out with ticket offices and a departure lounge.

But no one considered the wind around Midtown Manhattan's skyscrapers or figured out how to keep an airship, tethered only at its nose, steady enough to allow passengers to descend a narrow gangway into the mast. Releasing ballast for positioning required dumping gallons of water onto the streets below. As a result, the only thing ever delivered to the tower by airship was a bundle of newspapers—lowered, once only, by rope.

Man on Wire

Few skyscraper antics rival that of Philippe Petit, who walked on a tightrope between the towers of the World Trade Center in New York soon after their completion in 1974. His feat took years of planning—both to avoid building security and to figure out how to rig a line that would safely absorb the tall buildings' natural sway.

Petit used a bow and arrow to shoot a rope between the buildings. He then crossed it eight times in 45 minutes before surrendering to police. Sentenced only to performing for the children of New York, Petit is credited with bringing fame to two buildings that—until his walk—had hardly captivated the public.

Vertical Marathons

Each year in cities as far afield as Frankfurt, Seattle, Taipei, and Perth, hundreds of vertical races are held on the stairs of the world's tallest skyscrapers. Most of them raise money for medical charities—predominantly those working to combat cystic fibrosis, muscular dystrophy, or lung disease.

The speed at which runners tackle these stairs is impressive. The 86 floors, or 1,576 steps, of the Empire State Building were run in a record nine minutes and 33 seconds. Taipei 101's race was even longer—91 floors, 2,046 steps—and was completed in just about one minute more (10 minutes, 29 seconds).

F

façades, 150
in blast protection, 142
cleaning of, 148, 152–53
effect of weather on, 158–59
failure of, 158–59
light and heat transfer in, 70
ornamental, 48
types of, 63
windowless, 127
see also glass curtain wall
façades
face-sealed barrier walls, 68
face shields, 88
fall prevention and protection, 88
farming, vertical, 181
fatalities:
construction, 88
elevator, 100
from fires, 132, 138
from lightning, 128
Federal Building (Chicago), 36
Ferriss, Hugh, 176, 186
fiber optics, 119, 120–21
fire alarm systems, 134–35
fire control, 139
fire extinguishers, 144
firefighter lifts, 30, 136, 146
firefighting, 30, 108, 132–47
communications in,
124–25, 146
equipment for, 146
fire prevention, 144–45
fire pumps, 145
fire-resistance, 54, 75
testing for, 140–41

A Subway Rising

*Construction sites in
New York are typically
frequented by what are
jokingly called "roach
coaches"—small wagons
owned by vendors of
everything from coffee
to kabobs. But the site
of the new 1 World Trade
Center has boasted what
is perhaps the largest—
and certainly the most
upwardly mobile—
temporary food franchise
in history.*

*To save its workers
the time-consuming
descent by hoist to the
street at lunchtime, the
company erecting the
tower's steel opened a
Subway sandwich shop
on the twenty-seventh
floor in June 2010. Made
out of cargo containers,
it sits astride two tower
cranes and relies on
hydraulic legs to rise as
the building's skeleton
grows. The shop has
been open to any of the
1,000 or so laborers, of
all trades working on
the site.*

The Human Spider

Known as "the French Spider-Man" or "the Human Spider," Alain Robert is famous for scaling skyscrapers using only his bare hands and climbing shoes. He typically relies on window ledges and frames protruding from towers, and climbs early in the morning—before authorities are tipped off.

He has scaled many of the world's tallest buildings, including Taipei 101 (just before its opening in 2004) and the Petronas Towers (three times), as well as the Jin Mao Building (wearing a Spider-Man suit)—not always with ease. While climbing the Willis (Sears) Tower in Chicago in 1999, the top of the 110-story building disappeared in fog, making the last 20 stories a wet and unexpectedly hazardous part of the adventure.

Can You Name This Tune?

Sometimes, skyscrapers whistle. The CitySpire Center, one of New York's tallest mixed-use towers, annoyed its neighbors so much that workers had to remove every other louver in the cooling tower atop the building to quiet the wind moving through it.

Manchester, England's tallest skyscraper—the 47-story residential Beetham Tower—has whistled since it opened in 2006. First, foam pads were fitted to a thin glass blade atop the tower; in a second attempt, the blade was fitted with aluminum nosing. But the wind, and the noise, persisted—and efforts to quiet the building were renewed in early 2010.

Cemeteries in the Sky?

The world's tallest cemetery is 32 floors high. Located in Santos, Brazil, the Memorial Necropole Ecumenica III features more than just burial spaces: it has a restaurant, chapel, lagoon, and peacock garden as well. And it is open to people (dead ones, of course) of all religions.

Brazil has almost a dozen of these structures above 12 floors, many of them the product of a scarcity of land in heavily populated cities. But it's not alone in the vertical internment business: Bogotá, Colombia, completed a 13-tower mausoleum complex in 2006 and similar proposals have been floated for buildings in Hong Kong.

Numbering Madness

The number 4 in the Chinese language sounds like the word for death, and thus carries a negative connotation. As a result, some skyscrapers in China simply skip floors 40 to 49 (as well as 14, 24, 34, etc.), moving directly from 39 to 50 or even higher.

Henderson Land's recently completed 39 Conduit Road condominium development in Hong Kong goes directly from 39 to 60 (6 is considered lucky). The top three floors are labeled, consecutively, 66, 68, and 88. And just to make sure to cater to western tastes, the developer has carefully omitted a thirteenth floor as well—going right from 12 to 14.

"Skyscraper" Etymology

1788
The first reference refers to the name of a horse that won the Epsom Derby.

1794
A skyscraper is defined as a "triangular sky-sail."

1800
A tall hat or bonnet is referred to as a "skyscraper."

1826
The term "skyscraper" is used to define a tall horse.

1857
A very tall man is referred to as a "skyscraper."

1866
Fly balls in the newly minted games of baseball and cricket are called "skyscrapers."

1891
The word first appears to refer to tall buildings, particularly those in Chicago.

Prefab Skyscrapers

It's not your grandparents' camper. "Prefab," or modular, buildings consist of prefabricated modules that are manufactured in a remote facility and then delivered and assembled at a development site. Advantages of modular building include faster construction time, reduced costs due to on-site labor savings, and sustainable processes that produce less construction waste.

Victoria Hall, a 25-story apartment tower at the University of Wolverhampton in England, opened in 2009 as the tallest modular skyscraper built to date. The Atlantic Yards project in Brooklyn, New York, is considering a 34-story modular residential tower that may cut construction costs in half. Mobile home, indeed.